BACK AT THE FARM

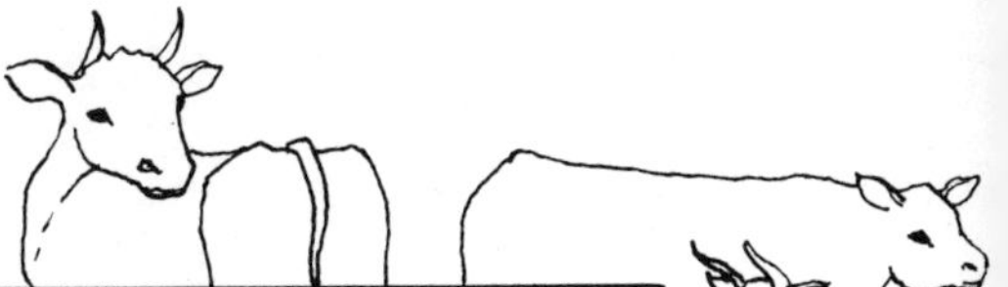

BACK AT THE FARM

Raising Livestock on a Small Scale

BARBARA AND DICK DEMING

VAN NOSTRAND REINHOLD COMPANY
NEW YORK CINCINNATI TORONTO LONDON MELBOURNE

Library of Congress Catalog Card Number 81-16479
ISBN 0-442-26334-1

Printed in the United States of America

Designed by Mildred Phillips

Published by Van Nostrand Reinhold Company Inc.
135 West 50th Street
New York, NY 10020

Van Nostrand Reinhold Publishing
1410 Birchmount Road
Scarborough, Ontario M1P 2E7, Canada

Van Nostrand Reinhold Australia Pty. Ltd.
17 Queen Street
Mitcham, Victoria 3132, Australia

Van Nostrand Reinhold Company Limited
Molly Millars Lane
Wokingham, Berkshire, England

16 15 14 13 12 11 10 9 8 7 6 5 4 3 2 1

Library of Congress Cataloging in Publication Data
Deming, Dick, 1935-
Back at the farm.

Bibliography: p.
Includes index.
1. Livestock. I. Deming, Barbara, 1941-
II. Title.
SF65.2.D45 636 81-16479
ISBN O-442-26334-1 AACR2

CONTENTS

PREFACE

FOR ALMOST two decades there has been a growing movement "back to the soil." The number of people who have taken up homesteading—that is, raising crops and animals for their own consumption—has greatly increased. If a surplus is produced, they sell their produce on a small-scale basis. The desire to homestead generally comes from two sources: an appreciation of the farming life and a deep concern for the way food is grown.

A great many of us homesteaders were born and raised in the city. Although our knowledge of farming may be limited, our commitment is firm. We have found frequently that a mere desire to homestead is not sufficient, and we therefore have devoted a great deal of time and effort to accumulating necessary knowledge.

For obvious reasons, city folk know more about gardening than about farm animals. But on a homestead, the animals seem to arrive . . . frequently two by two. The family, particularly children, greet each animal with enthusiasm, but too often without sufficient knowledge of how to care for it. It is to these homesteaders that this book is addressed.

We wish to express our appreciation to those people whose experience and support made it possible for us to carry on our work. Although there is a long list of these friends, we particularly want to acknowledge Vivian and Bruce Ackerman, Ruth Ernst, Betty Fitzpatrick, Carl and Gail Goodall, George and Virginia Hatch, John and Ethel Morgan, and Pete and Sue Scribner. We also would like to express our gratitude to some nonhomesteading friends: Ben and Diane Duce, John and Ruth Knowles, Mary and Cliff Gilson; and to our families, who have enthusiastically supported our life-style. A very special thanks to Mac Fiske for his dreams and fantasies that proved inspirational, and to Nap Barber, who has continuously assisted with the knowledge to pull them off. Special acknowledgment goes to Bill Connors, who repeatedly said, "You ought to write a book," until we did. Also, we thank Laura Pearlman, Marlene Cusner, and Virginia Hatch, our manuscript typists.

INTRODUCTION

TO THOSE of you who are concerned about the cost and quality of food, we offer this book to inform and inspire you to raise it yourself. We have primarily focused on the raising of livestock for meat, dairy products, and eggs because we found no single book that covered this topic sufficiently. There are many good books on how to raise vegetables, but few that begin with the basics of how to select, care for and profitably raise livestock on a small scale.

The ideal setting for this type of venture is a small family farm consisting of one to a few acres, but some of the animals discussed can also be raised in town with a minimum of yard space. For those who live in a city or suburb, we have included a chapter on the "postage-stamp homestead"; for those who already live in the country, we offer the rest of the book as a step-by-step guide to raising the various livestock you might consider.

The information is designed for homesteaders—people who raise food for their own consumption—rather than prospective commercial farmers. We encourage raising some surplus, but only to cover the cost of the family's food production. Homesteading is an avocation, not a vocation, and success depends in part on defining limits.

Homesteading is also a way of life designed to promote self-sufficiency, responsibility, knowledge of the life cycle, humor, and humility. A commitment by the entire family to homesteading is required to carry you through the discouraging times and to keep you from overextending during the good times. Aside from the pleasure of producing your own high-quality food at a low cost, you will discover that homesteading instills many personal values: children know their worth when their chores are essential to the family's basic needs, and everyone's sense of confidence and responsibility is nutured on a small family farm.

The primary resource needed to start raising livestock is a love and appreciation for animals. To be overly attached to them will work against you at slaughter time, but an appreciation of their worth as a food source can keep your sentiments in perspective. A trip to the butcher shop, where lamb chops sell for $4.50 a pound—as opposed to the $1.69 a pound you can raise it for—helps.

We have addressed both basic and advanced questions that are asked by people both before and during the raising of homestead animals. Types of animals and specific breeds within the class are discussed, as well as physical equipment required, work involved, medical problems, and realistic production goals. The cost of getting started with and raising a particular animal has been emphasized, since few people are in a financial position to raise animals without cost accountability, and it is an important part of good farm management. Some animals are less expensive to start raising than others, and we encourage prospective homesteaders to start with these and use the profits to expand to the more expensive ones.

Our knowledge has come from agricultural experts, books, magazines, farmers, other homesteaders and last, but not least, our own successes and failures over the eight years we have run the West-Hop Homestead.

BACK AT THE FARM

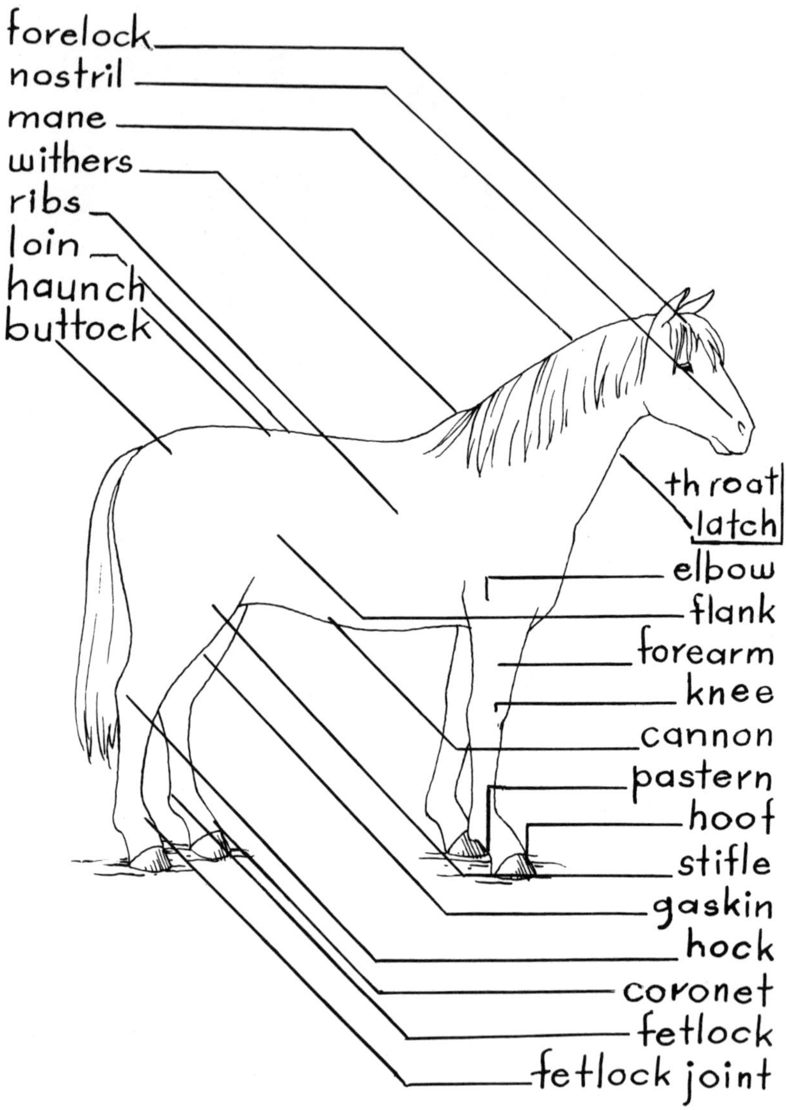
forelock
nostril
mane
withers
ribs
loin
haunch
buttock
throat latch
elbow
flank
forearm
knee
cannon
pastern
hoof
stifle
gaskin
hock
coronet
fetlock
fetlock joint

1 HORSES

THE WHOLE idea of moving to a little place in the country was to have room for a horse. We moved to the farm in mid-September, and three days later Lahneen arrived—a pinto grade horse about fourteen years old.

My dream was now complete. I envisioned sitting under the willow tree in the backyard, watching Lahneen nearby in her paddock; a good book and perhaps some music would complete the scene. But the rest of my family felt Lahneen needed company, and so we acquired Tequilla soon after. She was a four-year-old half-Arabian, a beautiful, spirited animal with a maddening ability to defy fencing.

As our first country spring rolled around, and everything started to blossom out, we discovered that Lahneen had been bred before we bought her. A horse's gestation period is eleven months, and after talking to her former owner we managed to narrow her delivery date to some time in July or August. For six weeks the entire neighborhood, including dogs and cats, checked on her daily. Her foaling in August remains the single most exciting event that has occurred on our homestead. We named her palomino colt Reprieve.

About this same time, we got Freddy, a Shetland pony, as more company for Tequilla. Freddy had more personality than any other horse of my acquaintance. His sense of humor was not always appreciated, but he certainly provided a lot of laughs. One of his tricks—my favorite—was to get an empty pail and wave it in the air. Tequilla came running every time, believing she was about to be fed.

In the seven years since then, a steady stream of horses and ponies has arrived and departed, with only Reprieve a permanent fixture. All the horses that have left us have gone to good homes, and all the new horses have been saved from the rendering plant. We have had up to five horses at one time, in addition to all our other animals.

With all that horsepower standing around, we decided to put them to harness. All of them, horses and ponies, have done their share of pulling logs for the wood stove and preparing land for planting. It is hard to justify a tractor on a five-acre homestead, so it makes good sense to put your horses to work.

TYPES OF HORSES

There are three general types of horses: ponies, light horses (light breeds), and draft horses (heavy breeds). Within each type, there are different breeds; a grade horse is one without registration papers, and can be either purebred or a combination of breeds.

A pony is any horse other than an Arabian that is shorter than 14.2 hands (56.8 inches) from the withers (the high point at the base of the neck) to the ground.

The most widely available pony, and the smallest, is the Shetland, which averages ten hands and weighs from 300 to 400 pounds. It may be solid-colored or spotted. A Shetland pony is often bought for children, who quickly outgrow it and may then neglect it. A pony of this size should not be expected to carry much more than 80 pounds, although it can pull a greater weight in a cart.

A Welsh pony averages 12.5 hands and weighs about 500 pounds, but there has been a trend among breeders in favor of even larger sizes, to produce sturdy riding horses.

The largest classification is the light horse; these weigh from 600 to 1,200 pounds and range from 14.2 to 17 hands high. Light-horse breeds include Arabian, Morgan, American quarter horse, American saddlebred, standardbred, and thoroughbred. Most homesteaders will not need a purebred horse; a combination (grade horse) is much less expensive, and often has a better temperament.

The draft horse, once an essential part of the rural scene, is making a comeback on small family farms. It usually is a very calm, which is one of its major attractions; and of course it is big and powerful. The three common breeds in this country are Belgian, Percheron, and Clydesdale.

This type of horse can measure well over 17 hands and weigh 1,400 pounds or more. They are capable of a lot of work; but considering their enormous appetites, they are not very practical for the small homestead.

If you think you need a work horse, a good compromise is a draft chunk, the result of breeding a draft stallion with a light mare. This horse costs less to maintain than a draft horse while providing more power than a typical light horse.

CHOOSING A HORSE

Before you buy a horse, determine how you will use it. If you plan to use the horse purely for pleasure, decide who will be doing the riding and/or driving. A pony can provide good service for children until they get too big to ride it, which is usually around twelve years old. However, ponies can also be cart-driven by adults, easily pulling 300 to 400 pounds in a cart. A pony can be broken to ride English or Western style, or broken to harness for draft work and driving. Because it is small, a pony is easier than a light horse for young children to take care of, in terms of both general management and saddling or harnessing. The size and eating efficiency of a pony means lower feed costs than with a light horse. Ponies are easy keepers.

A light horse provides the same services as a pony, but can accommodate adults as riders. The feed bill for a light horse will be at least twice that of a pony; however, you can expect a light horse to do more work for you.

If you are experienced and have the time to break a horse, you have a wider choice of animals; otherwise, buy an animal that is properly trained for all its intended uses. Don't buy a horse broken only to an English saddle if you intend to have all Western tack. The ideal animal would be well-broken to either Western or English style, be able to drive in a cart, and do some draft work.

Always ride the animal before buying it. If you feel unsure of yourself, take an experienced friend, and let him or her test out the animal.

Observe the horse's temperament. You probably want a calm, gentle horse that can be trusted around children. If you plan to use the horse for garden work, you surely do not want a skittish horse that may tear up your plants or ruin your equipment.

The horse's gender is a help in determining its temperament, although there are exceptions to every rule. Nonetheless, I would rule out a stallion (male horse) as a good multipurpose horse. A stallion may be highly spirited and difficult to control when around a mare (female horse). A mare may be very gentle, but also may be skittish at times, especially when she is in heat. A gelding (castrated male) is probably the best choice, as it tends to be quiet and dependable. In our experience, geldings have far more pleasant personalities than either stallions or mares. That is important considering the hours you will spend with the animal.

Age is a factor to consider, too. A horse is not fully developed for work before three years of age; don't buy one younger than that unless you plan to train it yourself. A horse reaches its prime at about five years of age, and with good care, should be sound for a number of years: a twenty-year-old horse is usually considered old, but some horses are still working at thirty. Since you are probably looking for a horse you can use for a number of years, but one that has some experience, consider a horse between five and ten years of age.

In selecting a horse, first look over the entire horse for good conformation and general appearance. The animal should look alert and attractive, and should have a good coat. The animal's general form should demonstrate good carriage; it should be symmetrical, with all parts blended together. Avoid buying an animal that has any trouble breathing when exercised, such as wheezing or difficulty in forcing air out of the lungs.

Some specific things to look for are as follows:

HEAD

- Jaw—the teeth of the upper and lower jaws should meet evenly.
- Eyes—check for partial or complete blindness; check for signs of any inflammation (redness, swelling, or discharge)
- Nostrils—should be large.
- Ears—medium-sized and well-carried.

BODY

- Chest—should have good depth and breadth with well-spaced ribs.
- Back—from the withers to the loin should be strong, and proportional in length to the rest of the horse; back should not sag noticeably.
- Rear quarters—should be deep and muscular.

LEGS and FEET

- Legs—knees should be centered properly: neither bent forward, backward, nor bowed; check for good action, with each foot carrying forward in a straight line; the step should be long, balanced, and easy.
- Feet—should be in proportion to size of the horse and in line with the legs; there should be no inflammation in the bottom of the hoof, and the hooves should not be split.

STABLES

Horses do not need elaborate housing, but they do need to be protected from severe weather, dampness, and drafts. Good ventilation is essential for all animals; horses are no exception.

BACK AT THE FARM

The size of the structure depends upon how many horses you have, of course, but ceilings should be at least seven to nine feet high. High ceilings allow more air circulation and prevent horses from bumping their heads.

The floor should be constructed of clay or wood. The floor should be kept as dry as possible; building it with a slight slant, particularly with a tie stall, will help the waste matter flow away from the horse. The harder the surface of the floor, the deeper you should pile the bedding for your horse's comfort.

Horses enjoy and need lots of light, but glass windows must be protected from breakage by the horse. Light bulbs and light fixtures, too, should be protected. These items may be protected by bars, heavy screens, or metal guards. Electric cords should not hang where the horses can chew on them.

The ideal arrangement is a box stall with a door leading to the paddock (exercise area) or pasture. This way the horse can wander, but has free access to the stall in inclement weather. If a door is impossible, a window will provide some light and ventilation.

A box stall should be at least 12 by 12 feet for a horse, and 8 by 8 feet for a pony. This size gives the horse enough room to move around and to lie down to sleep. Because there is a large area of bedding, it does not need to be cleaned every day.

If your stable is not big enough for box stalls, you can use tie stalls. A tie stall ideally should be at least 8 to 9 feet long and 4 to $4\frac{1}{2}$ feet wide for a horse, and 6 to 7 feet long and 3 to 4 feet wide for a pony. The advantage of the tie stall is that it requires less space, less bedding, and is easier to clean; but it must be cleaned daily. The stall should be on a 6- to 8-inch platform. The horse stands on the platform and leaves its droppings in the aisle below.

Horses can be satisfactorily housed in run-in sheds closed on three sides with the opening on the south (or whichever direction gets the least wind). This type of shelter provides maximum exercise.

There are a large number of bedding materials; your choice depends on availability, price, absorptive capacity, and potential value as a fertilizer. Peat moss has the highest rate of absorption, but it is expensive. Shavings and sawdust also work well, but they do not decompose as fast as the other materials. If you have time to gather them, leaves and pine needles make excellent bedding. They should be dry before being used. Bedding materials should not cause flying dust or be excessively coarse, such as wood chips.

Cribbing means chewing on boards, and most horses do it with varying degrees of frequency, generally when they are bored. To stop horses from cribbing their stalls, barn, and paddock railings, we periodically paint the wood with creosote or a solution of two-thirds motor oil and one-third turpentine. They will not eat this because of its terrible taste.

PADDOCK

The paddock, a fenced exercise and grazing area, should be near the shelter—not for the horse's sake, but so you won't get tired leading your horse back and forth. The ideal setup is to have the area attached to the barn so that the animals have a run-in in case of inclement weather.

There is no set size for a paddock, but if you do not have a pasture, the bigger the paddock is, the better. If you live in a section of the country where snow piles up in the winter, the horses will keep the area packed down, so they can exercise all winter.

Have the paddock ground slope away from the barn if possible; it will keep the ground in better condition after heavy rains. If the horses have access to the barn, the opening should be on the southern side, to protect the area from adverse weather.

The paddock should have a shaded area to provide all your horses relief from the sun when it is hot. However, if the shade is provided by leaf trees, the leaves cannot be left in the paddock in the fall. Damp leaves can become moldy, and if eaten could make your animals very sick.

The fencing around a paddock needs to be substantial. No matter what the size of the paddock, horses get bored and begin thinking that the green grass on the outside looks irresistible; and they break out of the paddock. A fence made out of heavy lumber is the best choice, especially if you use a strand of electric wire above the top board and perhaps a strand or two along the inside of the fence. The wire over the top board will prevent the horses from leaning over the fence, and from cribbing.

PASTURE

If you have enough extra land, you can pasture your horses. If the animal is relatively idle and the quality of grass is high, the summer pasture will provide sufficient food for the animal. Winter grass is not high in nutrients, so good hay or grain will have to be added to the diet. Also, check with your agricultural extension agent to find out if any plants toxic to horses grow in your area; make sure your pasture is free of them.

Early spring pasture can be highly laxative, especially for heavily worked animals. Wait until the grass matures, and then gradually accustom the animals to it. Each horse needs one to two acres of pasture for three to six months' use. Pasture areas should be rotated every year to minimize parasite infection; the parasite population increases as the manure accumulates. Horses do not eat weeds, so if you do not also have cows or goats, the pasture may need periodic mowing.

FEEDING

A horse owner has a variety of feeding programs from which to choose. The important thing is to choose one and stick to it. If you find you must change, do so gradually. The decision on types of feed should be based on cost, efficiency, and available storage space. You may select commercial feed, commercial feed and hay, or grain and hay to provide the protein, minerals, and vitamins horses need in their daily diet.

Horses can be fed hay alone if it is of good quality and the animals are not worked too hard. When buying hay, look for the following qualities:

- early cut hay from well-fertilized and limed land
- fine stems, neither woody nor stiff
- green color
- many leaves (USDA grade 1 or 2 hay is 25 to 40 percent leaves)
- no dust or mold (never feed dusty or moldy hay to horses)
- pleasant smell
- few weeds and foreign material

It is ideal to buy the hay right from the field at harvest time: you know the field conditions and when it was cut, and it will undoubtedly be cheaper. Be absolutely certain the hay you buy is cured properly, and therefore dry. Baled damp hay will mold, and wet hay stored loose is a fire hazard.

Legume hays, unless very plentiful in your area, are usually limited to one-quarter to one-third of the ration fed because of the expense. If you have good-quality legume hay, you can feed much less grain. Legume hays are apt to be dusty and/or moldy.

The types of legume hays are alfalfa hay, clover hay, and lespedeza. Alfalfa hay, when properly cured, is the most nutritious of the legumes.

Grass hays are lower in protein, calcium, and vitamins, but they are not as apt to be moldy and dusty. Generally, grass hay is cheaper than legume hay. A combination of grass and legume hays is excellent.

The types of grass hay are timothy, prairie hay, brome grass hay, orchard grass, and cereal hays (oat, barley, wheat, and rye). Timothy is the most popular hay for horses because it is relatively easy to grow, widely available, and they like it. It is low in protein, so if just timothy hay is fed as roughage, a high-protein grain such as oats should be fed with it.

Commercial feed companies offer a complete pelleted feed. These pellets contain hay for roughage, grain, vitamins, and minerals. This is all you have to feed. The advantages of the complete feeds are ease for the owner, less storage space, and less waste. The disadvantages are that the feed is costly and horses tend to crib more.

You can also buy commercial pellet feeds that are intended as supplements to hay or pasture.

Oats are the most popular grain for horses. Oats have a high fiber content, and are not likely to cause digestive trouble. As oats are higher in protein than most grains, they compensate for a low-protein grass hay. Oats can be fed as the only grain if the hay is half legumes. Oats have approximately 65 percent total digestible nutrients.

Corn is a very good horse feed, but it must be fed more carefully than oats. It has a high energy and low fiber content, and could cause colic if overfed. Corn has approximately 80 percent total digestible nutrients. Corn is high in starch and therefore provides heat for the horses in the winter; but it may cause the horses to overheat in the summer.

Barley is a good feed, if it is ground or soaked before feeding. Feed barley as a corn substitute in the same proportions as corn. Barley has approximately 77 percent total digestible nutrients.

Bran is a supplementary grain; it is palatable and a good source of the B vitamins which horses need. It is slightly laxative, however, so do not use it as the only grain. Wheat bran has a total digestible nutrient level of 66 percent.

The amount to feed your horse depends on the animal's weight and the amount of exercise it gets. Light work is considered to be two to three hours per day of exercise or work; medium work, four to five hours per day; heavy work, five to eight hours. A horse used under two hours per day is considered idle. Growing or lactating horses have a higher protein need and require different rations.

A general guide for daily rations is as follows:

- Horses used for light work should be allowed about $\frac{1}{2}$ pound of grain and $1\frac{1}{4}$ to $1\frac{1}{2}$ pounds of hay per 100 pounds of horse.
- Horses used for medium work should be allowed about 1 pound of grain and 1 to $1\frac{1}{4}$ pounds of hay per 100 pounds of horse.
- Horses used for hard work should be allowed about $1\frac{1}{4}$ to $1\frac{1}{3}$ pounds of grain and 1 pound of hay per 100 pounds of horse.
- Commercial horse pellets should be fed at approximately $1\frac{1}{2}$ pounds per 100 pounds of horse.

If you are mixing your own grains, some suggested proportions of grains are as follows:

RATION 1	RATION 2	RATION 3
Oats 100 percent	Oats 70 percent	Oats 70 percent
	Corn 30 percent	Barley 25 percent
		Bran 5 percent

Horses need to be fed two or three times a day. All the grain of a feeding should be eaten before the hay is given. We like to feed the hay ration more heavily at night, since it takes longer to digest. The rule of thumb is to increase the amount of grain and decrease the amount of hay as the animal increases its exercise rate. If the horse is used less than normal, the rule is reversed. Never feed grain to a horse that has not cooled down completely after working. Also, allow the animal about an hour to digest the food before working.

FEED COSTS

To determine the cost of feeding a horse, you first need to know its approximate weight. Next find out the price of the various commercial feeds and grains in your area. If you have a 500-pound pony fed light-work rations, your projected approximate costs are as follows:

FEED COSTS

Plan 1	*Pellets*
7.5 lb. × 30 days = 225 lb./month	
	@ $10.50 per 100 lb. of feed = $23.62/month
Plan 2	*Oats/Corn/Hay*
1.75 lb. oats × 30 days =	52.5 lb.
0.75 lb. corn × 30 days =	22.5 lb.
	75 lb. grain @ $9 per 100 lb. = $6.75/month
7.5 lb hay × 30 days =	22.5 lb. @ $2.00/35 lb. bale = $13.00/month
	$19.75/month

These prices and rations are only general estimates. Horses are like people in that their individual feed requirements can vary considerably from the estimated amounts. Adjust the feed rations according to how the animal gains, loses, or maintains its weight. The prices of pellets, hay, and grain also vary depending on where and from whom you buy your supplies. In our area, good legume hay can frequently be purchased from the farmer's field for $2 a bale in the summer, compared to $3.50 a bale or higher for low-quality grass hay in the late winter months. It really pays to have storage space.

All feed must be kept dry. Grain and pellets should be stored in covered, rodent-proof containers. The lids should be fastened so that the horse cannot get into the feed if it happens to get loose in the barn. Overeating of grain can cause a bad case of colic, or worse.

Horses need two to three ounces of salt per week. Commercial feeds contain salt, but it may not be in sufficient amounts for your animal. No matter what feed plan you use, always keep a salt block accessible to your horses.

Horses need five to twelve gallons of water daily, so devise some system to provide a constant supply of fresh water. An old bathtub in the paddock or pasture works well. Do not let a horse drink heavily when it is hot or just before you work it. Also avoid heavy watering after a grain feeding.

GROOMING

Grooming is necessary to keep the horse's coat in good condition and to help prevent skin diseases and parasites. You should do it every day, before and after exercising your horse.

For both ease and comfort, use a cross-tie to hold the horse. A cross-tie is a line from each side of the halter to a hook or ring set level with the horse's head. The cross-tie lines should be a permanent fixture in the barn.

To groom a horse, you need a towel, a currycomb, and a brush. Use the currycomb in a circular motion; start at the withers, working forward to the head, and then work from the withers to the rear. Do not stand directly behind the horse. Groom rapidly and vigorously, but not so roughly that you hurt the animal. Use a rubber currycomb on the face and legs, as the skin is thin and a metal currycomb would hurt.

When you start, place your free hand on the horse's shoulder to steady and reassure it. Talk to your animal. Horses love to be groomed, and enjoy your company.

Once the dirt has been loosened by the currycomb, use the brush to clean it off. Start at the back of the head and work forward to clean around the face, stroking down. Once the head is clean, brush the rest of the horse, always using strokes that follow the direction of the hair. When you brush the animal's legs, steady them with your free hand.

Brush the mane and tail last, working on them one section at a time. Finish by wiping the horse with a clean towel. There will be a nice sheen to the coat if you have done a good job.

To groom a horse after exercising, wait until it has cooled down. Walk it around quietly with the saddle girth loosened until the horse has stopped sweating. You may take the saddle off, but put a blanket over the horse immediately. Once it has stopped sweating, rub it down with burlap or a towel. A sweat scraper is handy to use before the cloth. When the horse has been thoroughly dried off by rubbing, use the currycomb and brush as described.

FOOT CARE

A horse is only as good as its feet and legs, so you must pay special attention to these areas. Check and clean your horse's feet before grooming.

Once you have the horse in the cross-tie, make sure its feet are well-placed under its body. You and/or the horse can get hurt if it is not in a secure position to be steady on three legs. Start by examining the front legs, so it knows what you are up to. Always face toward the rear of the horse when examining any of the legs or feet.

To lift a leg, place your free hand well above the leg you are going to examine. Talk to the horse while doing so to reassure it. Then run your working hand down the front of the leg to the cannon bone (the long bone below the knee). Grasp the cannon bone and at the same time press your free hand against the horse's shoulder. It will then shift its weight off the leg and, as it does, you can lift the leg and grasp the pastern (the bone below the next joint) with your free hand.

Turn the hoof up slightly so you can see the bottom of it to work on; support the horse's leg with your leg. If you need both hands to work, you can bring the leg between your legs and clamp it just above your knees. Keep hold of the pastern. Do not pull the feet back too far, and do not pull the front legs higher than the knees or the back legs higher than the hocks.

Once in position, clean the whole hoof, heel to toe, with a hoof pick. Clean out the packed manure, stones, and all other dirt. Watch out for signs of thrush (degeneration), bruises, swelling, cuts, or anything else that should not be there (see section on medical concerns, page 15).

Every four to six weeks, you will need to trim your horse's hooves. Trimming is not difficult, but it is important to maintain the proper angle. Have an experienced friend help you the first few times.

The hoof has three main parts to it: the wall, the sole, and the frog. The wall is the tough outer covering; it is thicker at the toe and gradually becomes thinner as it goes back to the heel. The sole is the bottom of the hoof just inside the wall, beyond the white-line separation. The frog is the horny triangular pad found in the center of the heel section.

To trim the hoof, use a pair of hoof nippers. Trim the wall to the level of the sole. The frog may be trimmed carefully by removing just the ragged edges. If you need to trim the sole, do it very sparingly. If you have to use a rasp, do not use it on the outside of the wall; it removes the natural hoof varnish that keeps the hoof from drying out. However, the sharp edge of the outside of the wall can be rounded a bit with the rasp after trimming.

If you are going to use your horse on hard surfaces for any length of time, shoeing is essential. Horseshoes can protect the hooves from cracks and corns and can correct poor hoof structure or growth.

Shoeing is a job for the farrier. A farrier has had training in fitting the shoes to the horse, corrective shoeing, and proper hoof trimming. The shoes are replaced or reset every four to eight weeks.

MEDICAL CONCERNS

Health problems can be kept at a minimum by proper care and management. The healthiest and best-cared-for animals get sick at times, however, and then you will need the services of a veterinarian. It is far easier if you already have an established relationship with a veterinarian before you need emergency services. If you need to call the veterinarian in an emergency, state the horse's symptoms and temperature. The normal temperature for a horse is 100° F (38° C). A temperature over 103° F (40°C) indicates a definite problem, and a temperature of 105° F (42° C) or more means the horse is very sick.

Preventive medicine is an important part of good maintenance. Annual vaccinations for equine encephalomyelitis and tetanus should be a routine part of your horse management. April is the best time to give these vaccinations, and many horse owners give the shots themselves. Be sure you take proper care of the serum and syringe, and that you give the injection correctly. Your veterinarian can teach you the procedure, and the serum can be bought at veterinarian supply stores. You may need to acquire a permit from your local health department to purchase syringes.

There are some items that should be kept in your barn medicine chest for general maintenance and for first aid in emergencies. Make sure that all medications are out of reach of the children. Some useful items are the following:

- scissors (for trimming hair)
- clean cloths or towels
- sterile cotton and gauze bandages
- worming pills
- mineral oil
- topical antibiotic lotion or powder
- neat's-foot oil
- Epsom salts
- astringent powder
- *veterinary* thermometer (*one with a loop at the end to which you can tie a long string*)

If you suspect an infectious disease, isolate the sick animal if at all possible and be sure to use separate water and feed buckets.

A horse can recover from a broken leg. Call the veterinarian immediately. Keep the horse as quiet as possible, and do not move the animal unless absolutely necessary.

Colic is a bellyache, and is common in horses. It can be caused by overeating, or eating or drinking too much water when overheated, or eating moldy feed. It can also be caused by constipation or impaction. The symptoms are signs of discomfort such as pacing, sweating, breathing hard, and indecision about whether to lie down. Sometimes a horse will try to poke with its head or kick its belly. A horse with colic may roll or thrash, which can cause further damage, so try to prevent it. Do not place yourself between the horse and a wall, as you can easily be injured. Call the veterinarian for anything but the mildest case; colic can lead to the twisting of an internal organ as the animal rolls around. A torsioned organ is usually fatal.

If the horse is hot and sweated, put a blanket over it; then start to walk the animal. You may even trot the horse slowly for a short distance to help break up the gas bubble, if the animal is not too uncomfortable.

Even after the horse is comfortable, keep a close watch. Many a horse owner has slept in the barn to make sure the gas does not build up again. Feed only good hay until you are sure the animal is back to normal. If the problem was caused by bad feed or moldy leaves in the paddock, get rid of the source of the problem.

Make sure that there are no nails, machinery, or junk materials where your horse will be. If a cut does occur, wash off the blood and press a dry, sterile cloth against the wound for up to ten minutes to stop the bleeding. If the blood is spurting out, use more pressure and hold a cloth there until the veterinarian arrives. Spurting blood indicates a cut artery, and requires prompt professional attention.

Once the bleeding is stopped, trim any hair away from the area. Keep the wound cleaned daily with warm, soapy water. You may apply a topical antibiotic. When the wound has a scab, keep it soft with petroleum jelly or bag balm (petrolatum).

There are many reasons for a horse's going lame, and do not hesitate to call the veterinarian if necessary. If your horse suddenly goes lame, first check its feet. The horse may have picked up a stone or some other object that can be easily removed.

You should also check for cracked hooves. Many horses have dry, cracked hooves, and if the crack has been forced open, stones can work up into the crack. If your horse has this problem, keeping the ground wet around the water tank will help. You may also use neat's-foot oil, or a commercial product for hoof treatment.

The swelling of the tendon below the knee or hock at the back, called curb, can cause lameness. A bruising of muscles or tendons can also cause lameness. For these conditions, try soaking or compressing the area with hot Epsom salts and water. Liniment applied to the area should help also. If improvement is not seen immediately, call the veterinarian.

Founder is an inflammation of the sensitive part of the inside of the hoof, usually affecting the front feet. It is a very serious condition and is apt to impair the horse permanently. The symptoms are severe lameness, considerable warmth around the foot, and a definite pulsation in the lower leg. The problem is often caused by overeating of lush grass or excessive amount of grain. This condition definitely requires the assistance of a veterinarian.

Thrush is a disease of the frog of the hoof, and is almost always caused by unsanitary conditions. There is usually a discharge with a disagreeable odor. Thrush must be treated quickly with antiseptics, or it will lead to severe lameness.

Sprains are not uncommon for horses. Treat a sprain on a horse as you would for yourself. Use cold water or ice and alcohol if it is a fresh sprain, and hot packs and heating liniments if it has already swollen. Treat the sprain several times a day. Massaging the area helps as well.

Strangles or distemper (shipping fever) is a very contagious disease; if suspected, isolate the animal. The symptoms are a heavy mucous secretion from the nose, a cough, a high fever (104° F to 106° F), and swelling of the glands under and between the jawbones. This disease requires the assistance of a veterinarian.

There are three types of equine encephalomyelitis and all three are a virus-produced disease spread by mosquitoes and other biting insects. The three types are Eastern equine encephalomyelitis (EEE), Western equine encephalomyelitis (WEE), and Venezuelan equine encephalomyelitis (VEE). VEE has only occurred in Texas to date, so unless you live in or are taking your horse through the Southwest, just the EEE and WEE vaccinations are needed. The vaccine is effective for six months, so have the horse vaccinated in April each year.

The symptoms for all three forms of equine encephalomyelitis are lack of coordination, sleepy appearance, drooping lip, and high fever. Both EEE and VEE have as high as an 80 to 90 percent mortality rate; WEE has about a 50 percent mortality rate.

Tetanus (lockjaw) is caused by bacteria that usually enter the body through a wound. The first symptoms are a stiffness about the limbs, difficulty swallowing, or refusal to eat. Your horse should have a tetanus toxoid booster every year. If you know your horse has a puncture wound, an additional injection of tetanus antitoxin is a good idea. Once the disease has progressed to noticeable symptoms, it is almost always fatal.

Equine infectious anemia (EIA), also called swamp fever, is caused by a virus and is spread primarily by biting insects such as mosquitoes and horseflies. The symptoms are a high fever (104° F to 108° F), depression, loss of appetite, and rapid deterioration. There are three forms of the disease. The acute form often ends with the horse's dying within three to fourteen days. The subacute form is similar to the acute form, but the animal appears to recover for a period of time. The symptoms usually recur, particularly during times of stress. The third form is chronic, in which the animal is symptom-free.

The screening device for EIA is the Coggins test. The test involves drawing blood from a horse, and then testing the sample to see if it registers negative or positive to the agar immunodiffusion test. Most state regulations require that, if the test is positive, the horse be quarantined for life—which means isolated and never used again—or destroyed. Prior to the use of the Coggins test in 1972, fewer than 500 horses of a total horse population of 10 million in this country died each year from EIA. Since the Coggins test began to be used, more than 26,000 horses have been ordered to be destroyed or quarantined. It would seem that the test is more fatal than the disease.

There is considerable conflict between the law and the research done on EIA. One problem is that the Coggins test does not test for presence of the virus, but in fact tests for the antibody level in the blood. The second problem is that the great majority of horses testing positive are considered to have the chronic form, and research shows that the chronic horse is a nontransmitter. Proponents of the regulation claim that, although it is true that a chronic horse cannot transmit EIA, the disease may at some time change to the acute or subacute form. The research on EIA done by the New England Institute of Comparative Medicine shows this to be an invalid assumption, and implies that thousands of horses have been killed needlessly since 1972.

Our introduction to the problems of EIA came when we had Lahneen Coggins-tested six weeks prior to her foaling. The only reason we had her tested was that the veterinarian was there to give the annual vaccines (Lahneen had never shown symptoms of any disease). Representatives from the Division of Animal Health brought us the news that she had tested positive and gave us a choice of screened quarantine or destroying her in accordance with the state regulations.

We chose quarantine, and with the seventy-two hours allotted us we had constructed an unbelievably secure, screened quarters out of our garage. The literature we received at that time indicated that the mare's foal had a fifty percent chance of having EIA. It seemed a high-risk situation, but a necessary one because of the emotional involvement with the unborn foal felt by all the children in the neighborhood.

Dr. Kittleson from the New England Institute of Comparative Medicine contacted us when she learned about our situation, several weeks after Reprieve was born. She told us that the institute had done extensive research on horses with EIA parentage and that its findings demonstrated that even though both of Reprieve's parents were positive reactors to the Coggins test, he would be negative by the time he was one year old. Foals from a positive horse will be positive while nursing and for a period of time after weaning because the antibodies are passed to the foal through the milk.

All of our experiences supported the institute's research; Lahneen has never gone acute or subacute even after stress periods, and none of our other horses has ever tested positive to the Coggins test. One mighty cheer was heard throughout our neighborhead the day the one-year-old Reprieve's Coggins test was pronounced negative.

The only way you can be sure a horse is worm-free is to have a veterinarian do a fecal sample test. The veterinarian can also tell you what type of worms the horse has if worms are present. There are no reliable symptoms for low levels of infestation. High levels of infestation can cause death. There are worming medicines for the more than fifty varieties of worms that can affect horses. A few of the more common worms are described here.

Ascarids are long, white roundworms. They more commonly affect horses under five years old. Bots are larvae of the botfly. Good grooming helps prevent infestations of bots because the white eggs are laid and attached to the hair, especially on the legs. Horses will bite off these small eggs if you don't get them off first. Pinworms come in two varieties; one type is whitish in color, with a long tail; the other type can hardly be seen. A horse with pinworms will often rub its buttocks. Strongyles come in two sizes, large and small. These parasites travel through the bloodstream and can cause serious damage to horses.

Tapeworms of three types can attack horses. Clean bedding, removing the daily manure accumulation, and rotating pastures are good preventive measures.

There is no denying the high cost of a horse both in terms of time and money, but the cost can be more than compensated by the pleasure a horse can bring the whole family. The key to maximizing this pleasure is to realize the versatility of horses. Use of a horse can range from bareback riding through fields or woods to the gymkhana and show circuits, or from having the satisfaction of cultivating straight rows in the garden to piling the kids in the horse-drawn cart to visit neighbors. No other animal—or hobby—can provide such a wide range of a year-round satisfaction to an entire family.

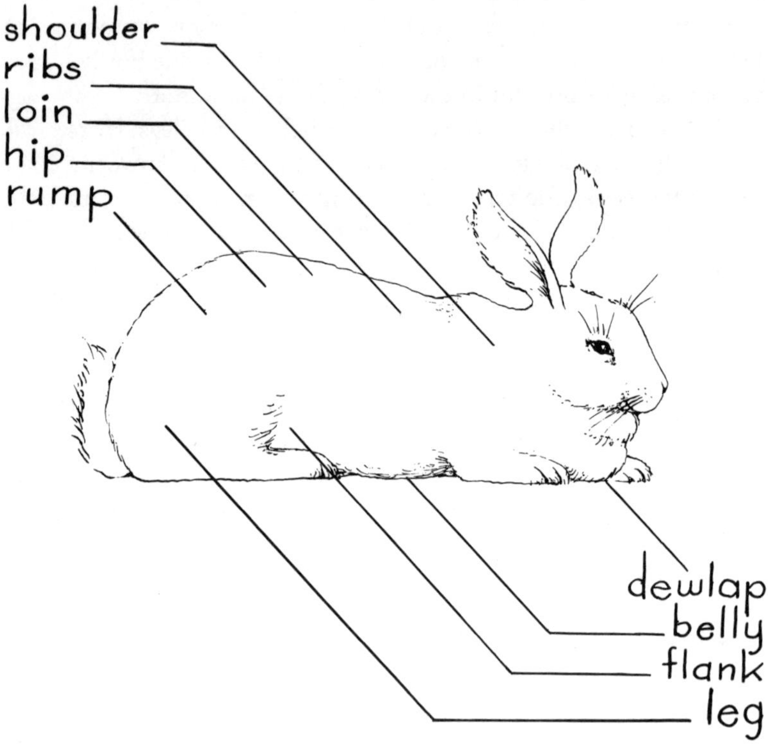
shoulder
ribs
loin
hip
rump
dewlap
belly
flank
leg

2
RABBITS

IN THE SPRING of our first year, we started thinking about raising some type of meat animal. Rabbits, we decided, would be perfect. The equipment needed is simple, and we thought the children could help with care and management. Everybody knows how prolific rabbits are, so raising them sounded easy. We figured one rabbit could be bred six times a year and produce an average of eight bunnies per litter. We would dispatch the young to the freezer when they weighed between 4 and 5 pounds, giving us between 192 and 240 pounds of rabbit per year from a single doe. What a machine!

Sold on the merits of rabbit raising, we set out to buy our herd. Good fortune brought us to Mr. Stevens's rabbitry. He had been raising rabbits for forty-seven years, and his New Zealands reflected his knowledge and experience. The fact that our whole herd survived our first year of management attests to their hardiness.

One of the first steps Mr. Stevens took was to acquaint us with some rabbit-raising terminology. A female rabbit is a doe, a male is called a buck, and they live in hutches. A junior is a rabbit under six months of age. A cull is a rabbit that has been picked out because of inferior quality; do not buy a cull for breeding stock. Giving birth is referred to as kindling, and a good doe will kindle not fewer than eight bunnies and not more than twelve at a time. With these terms in your vocabulary, you will be off to a good start in anybody's rabbitry. We learned also that rabbit breeders do not like to hear how "cute" their rabbits are; they see them as majestic and beautiful. If the children are with you when you visit a rabbitry, forewarn them not to make any loud noises or stick their fingers in the cages. Rabbits are extremely high-strung, and must not be subjected to any unnecessary stress.

As time passed in our rabbit raising, reality set in. Our youngest children are too spontaneous and noisily enthusiastic about life to work around such stress-sensitive animals. Our projected production figures were overly optimistic. We have found it to be more accurate to plan on successfully breeding a doe four times a year instead of six, which means one doe will produce around 130 pounds of rabbit per year. Then, of course, the important pounds are the edible ones, and rabbits dress out between 50 and 60 percent—which means that one doe provides you with between 65 and 78 pounds of edible meat.

A factor to keep in mind when raising meat animals is the feed: meat conversion ratio—the number of pounds of feed required to produce one pound of meat. For rabbits the conversion ratio should be 4:1 or less. This figure includes the feed needed for the doe and the buck as well as their offspring.

TYPES OF RABBITS

The rabbit has come a long way since it first was considered a food source in Asia more than three thousand years ago. Now there are more than fifty breeds to choose from, providing a choice among four different sizes, four different types of fur, and many different colors. A good way to see the different types of rabbits and to develop a feel for a well-bred rabbit is to visit a rabbit show or an agricultural or 4-H fair in your area.

Rabbits can be used for meat, fur, wool, laboratory experiments, fertilizer, pets, or a combination thereof. The best breed for you depends upon what you want to do with the rabbits. The white New Zealands are the most popular all-around breed, undoubtedly because they are so versatile. They are efficient meat producers, they are the only rabbits accepted for laboratory use, and their fur can be used in its natural color or dyed.

If your primary consideration is economical meat for yourself, choose a breed that appeals to you from the small or medium-sized rabbits. If you plan to sell some of your rabbits for meat, keep in mind the preferred breeds in your area.

SIZE

Among dwarf rabbits, which weigh full-grown from 2 to $3\frac{1}{2}$ pounds, the two most popular breeds are the Netherland Dwarf and the Polish. Dwarfs are usually bred for extremely small size and distinctive markings. They are not really suitable for meat production.

The breeds classified as small weigh four to seven pounds full-grown; examples are Dutch, English Spot, and Standard Chinchilla. These rabbits carry a lot of meat on their frames and require less space than the medium or giant breeds.

The mature medium-sized breeds reach nine to twelve pounds. This category includes the most popular breed of rabbit, the New Zealand, as well as the Californian, Satin, Champagne d'Argent, and the Lop. Lop rabbits have distinctive drooping ears, making them the basset hounds of the rabbit world. A medium-sized rabbit is a good choice for meat production; it can provide a fine-boned fryer of four pounds or more by weaning age.

Rabbits classed as giants weigh twelve to fifteen or more pounds and have reached a record weight of twenty-three pounds—that's a lot of rabbit! Included in this category are the Flemish Giant and the Checkered Giant. The giants' bones are large and they have large appetites, both of which factors negatively affect their feed:meat conversion ratio.

FUR

If you are interested in using or selling the hides, choose a breed according to the fur type you want. There are four types of fur. Normal fur is about an inch long and can be found in all weight groups. The New Zealand and the Dutch rabbits are examples of rabbits with normal fur.

Angora fur is about three inches long and can be spun into knitting yarns. The English and French Angoras are both medium-sized rabbits. There is not much of a market for the wool, but local handweavers may be interested in it. The wool should be plucked or clipped every three months. You should not take wool from a breeding doe; she needs it for her nesting box.

Satin fur is about an inch long, very dense, and has a greater color intensity and luster than normal fur. Satin rabbits are medium-sized and come in a variety of colors.

Rex fur is quite plush and short, with the hair only about five-eighths of an inch long. Rex rabbits are medium-sized and come in many colors.

Most homesteaders are interested primarily in rabbits for meat. For this purpose, the New Zealand, Californian, Satin, and Champagne d'Argents top the list of satisfactory utility breeds.

BREEDING STOCK

A call to your local agricultural extension office will get you a list of rabbit breeders in your area or the names of the people connected with the American Rabbit Breeders Association (ARBA).

Before you set out to buy your stock, you should be familiar with some breeding terms. A grade rabbit is any rabbit without papers. The grade may be a purebred, but without papers you cannot be guaranteed of that fact, nor can you prove it to any potential buyer of your stock.

Purebred rabbits are bred to a standard which is set by the ARBA. When you buy a purebred, you should get a paper showing the rabbit's pedigree. The pedigree number will match the number tattooed on your rabbit's ear.

Registered rabbits are full-grown rabbits that are purebred, papered, and have been examined by a licensed ARBA registrar and found to meet the minimum standards for the breed. A registered rabbit will have registration papers in addition to pedigree papers and will also have a registration number. These numbers should be tattooed on the rabbit's ear.

The advantage of papered stock is that you are guaranteed of having good stock and therefore predictably good offspring. It costs the same to feed a poorly bred rabbit as it does to feed a well-bred one, so why not go first class? We have also found that the purchase price for a papered rabbit usually is the same as for a grade rabbit; registered rabbits can cost considerably more and are usually not worth the expense if you will be using them strictly for meat.

When you buy breeding stock, be sure to tell the breeder your plans for the rabbits. Rabbit breeders take pride in seeing their line expanded, and they will help you choose the proper animals. There is apt to be a difference between the price for breeding stock and for meat, but if you pay meat prices you will not get papers with your rabbits, and you might get culls. Remember that the breeder feels the cull is inferior for some reason and is not worthy of raising as breeding stock. Your breeding stock should be the best of litters from the best breeding stock available.

CONFORMATION

Judging a good rabbit is impossible if you do not know what to look for, and that takes experience. In general, you want an animal that has good conformation; its parts should give a balanced look to the whole animal, so that it is proportional and pleasing to look at. Look for roundness, width, and fullness, especially in the hip area.

The rabbit should feel firm, not fat, and should have a nice coat. Avoid buying animals with sore hocks (sores on their hind legs), indications of ear troubles (sores or mites), or any sign of a cold.

OTHER CONSIDERATIONS

When buying breeding stock, check the breeding history of the animal and of its sire and dam (father and mother). Look for stock that is easily bred, produces litters of about eight in number, and has good weight gains. This information should be on hutch cards on the cage of each rabbit in the rabbitry. Breeders who are conscientious about their stock will have them.

The number of rabbits you start with needs to be decided. One buck can service ten does, but if they are all at breeding age when you buy them you could find yourself with eighty rabbits within six weeks of setting up. That may be more bunnies than you are ready to cope with.

Rabbits purchased prior to reaching maturity will be cheaper than a proven buck or doe (one that has produced at least one litter), but you will have to wait until they reach breeding age to get them into production. A doe will be maximumly productive for five to six years; a buck's productivity is usually around eight years.

Decide if you want to go commercial or if you are just raising meat for your own use. Two to four does and one or two bucks is a good start for the average family. Use the meat from one or two does and sell the litters from the others to pay for everybody's feed. In our area, live rabbits bring 90¢ a pound when sold directly to a customer, and 75¢ a pound from a dealer. For us, this means that one of our dinners and one or two of our lunches per week are free.

HUTCHES

Rabbits may be housed outdoors or indoors, as long as they have light and ventilation, are kept dry and out of winds and drafts.

If the hutches are outdoors, make sure the rabbits have ample shade in the summer and cover during inclement weather. Some breeders successfully use cages with wooden roofs and empty feed sacks on the sides for protection during heavy rain or winter. You can use wood for the sides, but then you have the problem of keeping

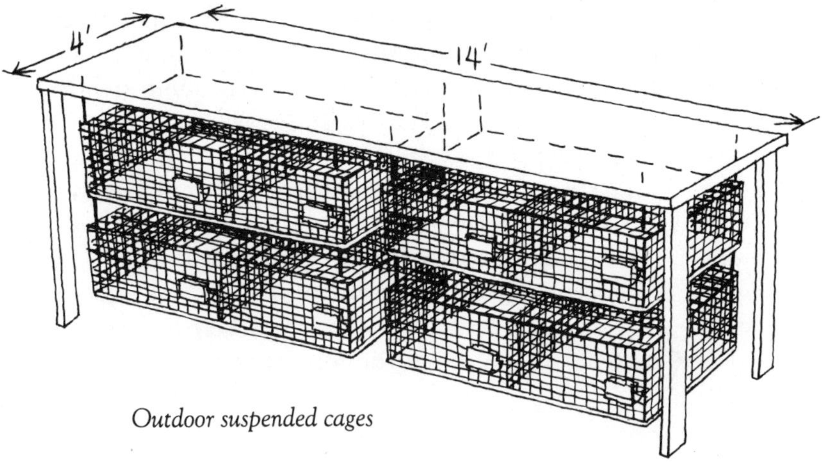

Outdoor suspended cages

the sides clean and keeping the rabbits from chewing the wood. A compromise system is to build a lean-to of wood from which wire cages are suspended. If the cages are on hooks, you can easily unhook them and remove them from the frame when it is time to clean them. The cages on the upper level of the arrangement must have some system to catch waste material.

The size of the cages depends upon the type of rabbits you have and whether the feeders and waterers are on the outside or inside of the cage. If the equipment is on the outside (the preferred location), the standard size for the medium-size breeds is thirty inches by thirty-six inches by eighteen inches high for the breeding doe and her litter. For the small breeds, the cage for the breeding doe and litter can be twenty-four inches by thirty inches by fifteen inches. The smaller cage is large enough for the mediums as a buck cage or a holding cage for bunnies after weaning. The giants need an additional square foot or two of space. All cages need a door opening at least a foot wide for your ease.

The cage material will help determine how successful and enjoyable rabbit-raising will be. An all-wire hutch is the best. It is self-cleaning, drier, more sanitary, allows you to observe the rabbits easily, is economical to construct, and lasts forever with minimum care. It promotes healthier and therefore more prolific rabbits, which, after all, is what rabbit raising is all about. You can buy precut caging from rabbit equipment suppliers or rabbit breeders, or you can build your own. The cheapest method is to buy from a rabbitry that is going out of business or reducing its stock.

You need galvanized, welded-wire fencing, either fourteen or sixteen gauge; the former is heavier and more expensive, and the sixteen gauge is adequate for the medium and small breeds. The floor should be 1-inch-by-$1\frac{1}{2}$-inch mesh; the top and sides 1-inch-by-2-inch mesh.

NESTING

Each doe needs a nesting box for her litter. If you want to spend the money, you can buy wire nest boxes with a disposable inner liner. The liners are made of cardboard with a waxed coating, and when the litter leaves the nest, you just dispose of the liner.

Homemade nesting boxes work just as well. The medium-sized rabbits need a nesting box at least eighteen inches long and ten inches high. We followed the open nesting box plan suggested in the *American Rabbit Breeders Association Official Guide to Raising Better Rabbits*. Some breeders prefer a partially closed nesting box, but this type of box tends to hold moisture in the nest.

Nesting boxes should be constructed of a light wood such as pine or plywood. Three-quarter-inch plywood is good for the ends, half-

Open nest box

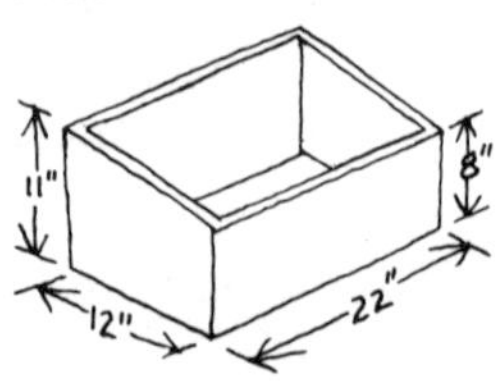

inch for the sides and bottom. If winter temperatures in your area frequently fall below 0° F, you might want to cut a second piece of wood to fit inside the bottom of the box. In the winter, put newspaper between the bottom and the second piece of wood for insulation. You may cover the edges of the nesting box with galvanized metal to keep the does from chewing the wood.

BREEDING

Rabbits are ready to breed when they are mature: in small breeds at about five months or five pounds; in medium breeds at about six months or ten pounds; and in giants at about eight months or thirteen pounds. They should be bred within six months of reaching maturity, as too long a wait can cause problems with fertility.

There is some disagreement as to whether does have a heat cycle. Some authorities believe that mature females are infertile for a period of two to four days every twelve days, and other authorities will tell you that they are always fertile. If you are having trouble getting a doe bred, shift the breeding schedule to allow for a period of infertility or low service receptivity. The doe does not ovulate until about ten hours after sexual stimulation; the eggs are receptive to fertilization for approximately six hours after ovulation.

When the buck deposits sperm, it takes about thirty minutes for the sperm to reach the other end of the female reproductive tract. If the doe urinates during that period, the semen could be washed away. Once the sperm reaches the oviduct, it takes six to eight hours for it to mature to the point of being able to fertilize an ovum. The sperm will live for about twenty-four hours in the female's reproductive tract; therefore, taking the doe back to the buck for a second servicing eight to ten hours after the first one increases the chances of conception.

Rabbits can be bred year-round, but the first six months of the year are usually the most productive. The heat of late summer can cause temporary sterility in a buck. Examine his testicles to make sure they are full, large, and descended into the scrotum before using him. Withered and withdrawn testicles indicate sterility, but if due to the heat, it will only be temporary. Also, soaring temperatures are a stress factor on your does, and may affect their reproductive ability.

The fall and early winter periods also cause breeding problems because of stress caused by changing weather. The shorter days sometimes affect animal breeding patterns. If you live where the winters are very severe, it might be to better advantage to keep the cages in a garage or other building. Newborn bunnies are very cold-sensitive.

Breeding Methods

There are four methods of breeding. Inbreeding is the breeding of close relatives, such as father-daughter or sister-brother. Linebreeding is the mating of less closely related animals, such as aunt-nephew, with the idea of retaining a close relationship to some excellent ancestor. Outcrossing is the mating of unrelated animals within the same breed. Outcrossing, particularly with an unknown line, can result in negative recessive characteristics becoming dominant. Crossbreeding is the mating of two different breeds, for example breeding a New Zealand with a Satin.

In the business of raising rabbits, inbreeding and linebreeding are the best methods if you have started with good linebred stock. By inbreeding, you ensure the continuation of the good characteristics found in a line. However, it also ensures the continuation of bad characteristics, so you must start with the best stock available.

Does are highly protective of their hutches, so always take the doe to the buck. Wait for him to service her and then take her back to her own cage. Never leave them alone. If you do, you will not know if she is bred, and if she tires of the buck's continuing advances, there could be a fight. Rabbits usually breed quickly if they are receptive.

The buck will mount the doe. The doe will have raised her hindquarters, and the buck will service her. It is over when the buck falls over on his side or falls backward.

Occasionally, things do not go smoothly. If the buck is not interested but the doe is, put the buck on her back to stir his instincts. If not, take the buck to the doe's cage and leave them in each other's cages over night. The next day take the doe back to her cage, let the buck service her, and then take him back to his own cage.

If the doe is not interested, try the cage-switch system. However, if the doe seems disinterested but her tail is twitching, give her a little more time to let the buck catch her. If you continue to have a problem with breeding, check on the following problems:

- Weight: are both the buck and the doe at ideal weight? Overweight bucks can get lazy, and overweight does have problems conceiving. Do not feed the rabbits until after you have bred them.
- General health: rabbits not in good health are not interested in breeding.
- Sterility: check that the buck's testicles are full and descended. If not, hope that the sterility is temporary.
- Check the does's vulva to see if it is reddish purple in color, which indicates she is ready to be bred.
- Stress factors: too little light, too much dampness, weather extremes, or inadequate food and water all can cause breeding problems.

You have a good breeding doe if she does the following:

- Can be bred easily: accepts service readily, and conception takes place.
- Eats well: poor eaters have trouble kindling and will lose more bunnies.
- Builds a good nest and puts the litter in it.
- Produces eight to twelve bunnies per litter and nurses well.

An extremely important part of supervising the breeding is to record the date the doe was bred and to which buck, if you have more than one. Also, record the kindling date, number in the litter, and the bunnies' development. You can buy hutch record cards from any rabbitry supplier, or use recipe cards. It is best to have the record attached to the cage or the feeder, if it is outside the cage, as a reminder to keep the records up-to-date.

If all goes well with the first servicing and you returned the doe to the buck again some eight to ten hours later, the doe should be bred. The only way you can be positive, short of waiting thirty-two days or so, is to palpate her to see if you can feel the young. About two weeks after mating, put the doe on a flat surface, holding her by the scuff of the neck with one hand, and slide your other under her belly. Feel very gently just forward of the groin area and slightly to the left and center of the belly. The fetuses will feel like large marbles.

Palpation can save you a lot of time if the doe is not bred, but it takes practice. The fetuses are hard to distinguish from internal organs, and some breeders feel that palpating can cause injuries to the fetuses.

KINDLING

The gestation period for rabbits is twenty-nine to thirty-four days, so put the nesting box in the doe's cage twenty-seven days after mating. Do not be overly efficient and give it to her any earlier; she will probably use it as an added bathroom and then refuse to kindle in it.

For a doe to have a successful kindling, she needs an adequate nesting box and a quiet environment. This last is essential. From the time you place the nesting box in the cage until the bunnies are four to five days old, make every effort to see that the doe is not agitated or frightened. Children, dogs, cats, and loud noises can be a cause of litter mortality among healthy rabbits. If the doe is frightened shortly before kindling, it can cause her to kindle prematurely; if agitated after kindling, she may destroy her litter.

If the weather is warm, put shavings, hay or straw, or commercial shredded sugar cane (available at farm-supply stores) in the nesting box for the doe to build her nest. Do not use sawdust, which can suffocate the bunnies and contribute to eye problems.

In cold weather, put a layer of newspaper in the bottom of the box, cover it with the wooden liner and then put in the bedding material. You can also use corrugated cardboard for additional insulation on the floor of the box. Some rabbit breeders recommend Styrofoam to line the nesting boxes in winter, but our rabbits chewed on it.

The doe will pull out hair from her underside to cover the bunnies. Some rabbits pull hair out before and after kindling, and some only afterwards. You might save some of the extra hair pulled in the warm weather for additional covering during the cold months, but make sure you do not give one doe another's hair.

BUNNIES

Once the litter has arrived, check the bunnies. Wait until the doe has calmed down a bit; with some does it is best to wait a day. Give the doe a treat such as half a carrot or a lettuce leaf to distract her, and then move the nesting box to a place where she cannot see you looking through it. Dab vanilla extract on her nose when you put the litter back. By the time the vanilla wears off, the box and litter will smell like her again and she will not destroy the bunnies.

Carefully push the bedding aside and count the bunnies, removing any dead ones. The ideal litter is eight, the number of nipples a doe has. If you have more than twelve, you may give the extras to another doe if you have had one kindle within a few days of this litter's birth. You can also cull by disposing of the runts. Some does can handle more than twelve bunnies, particularly if they are weaned in stages. But if it seems that the bunnies are not getting enough food, remove the smaller ones.

Keep checking the nesting box until the bunnies are ready to leave it. Remove any dead ones promptly and be sure the bunnies are all in one place in the box. A doe feeds only in one place, so if the litter is divided, only half will get fed. Watch for any bunnies outside the box, as occasionally a doe will get tired of nursing or become startled, and will leave the box with a sucking bunny hanging on.

Bunnies are born without fur, and with their eyes closed. A few days after birth, the fur starts to appear. In about ten days their eyes will open. Check their eyes to make sure they have opened and are free of conjunctivitis, which is much easier to treat if caught early.

The bunnies will leave the nesting box between fourteen and

twenty-one days. In the summer they tend to leave at the earlier time. Once they start to leave the nesting box, you may remove it, but before you do, wash down the hutch floor with a disinfectant solution.

Determining Gender

While checking the bunnies, you should determine their gender. This is easier said than done, but it gets easier as the bunnies get older. Place the bunny on its back with its head down, against your fingers. Use an index finger to press the tail back and down. Then gently push down on the sexual organ to expose the reddish mucous membrane. The mucous membrane of a buck will protrude sufficiently to form a circle. With a doe, the mucous membrane will form a slit that will have a slight depression at the end toward the rectum.

The sexing of young rabbits is quite important by weaning time because immature rabbits have been known to breed, with disastrous results to the doe and offspring. When you put them in holding cages at weaning time, divide them into separate cages by their gender.

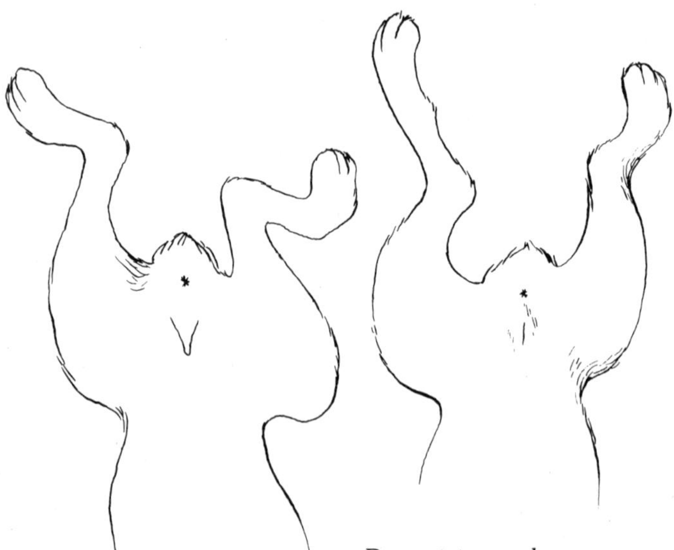

Determining gender

Weaning

The bunnies are ready to be weaned at about eight weeks of age. They should be eating pellets well, and you may have already rebred the doe when the litter was six weeks old. An excellent system for large litters is to wean the largest of the bunnies at six weeks and the rest at eight weeks. This not only allows the doe to dry up gradually, but also improves the weight of the smaller bunnies without any negative effects on the bunnies weaned at six weeks.

With the medium-sized breeds, you want to produce litters that are uniform in size and reach the following weight development goals:

21 days—average of $\frac{3}{4}$ pounds each
42 days—average of 3 pounds each
56 days—average of 4 pounds each

If you fall short of these goals, your feed:meat conversion ratio is going to be negatively affected. Good breeding stock, adequate litter size, and proper diet will produce these results with no difficulty.

FEED

The general protein requirements vary according to the status of the rabbit. Dry does and bucks need about 15 percent protein; pregnant and lactating does about 17 percent protein; and preweanling bunnies about 18 percent protein. Crude fat and crude fiber requirements run about 2 to 3.5 percent, and 15 to 27 percent, respectively. If you are feeding pellets, check the ingredient tags before buying. The feed for pregnant and lactating does may be fed to everyone to simplify matters, but you will be spending a bit more money than you need to.

If you feed pellets in the proper rations and provide all the fresh water your rabbits want, their nutritional needs will be met. Rabbits' individual feed needs vary, so use their weights as a guide to feeding. You need to check their weight regularly to verify both proper rations and peak condition for breeding. Fat rabbits do not breed easily, and sometimes not at all.

The general daily rations by breed size are: small breeds—at least two ounces; medium breeds—five ounces; giant breeds—eight ounces. A doe with a litter should be free-fed. Some people free-feed all of their rabbits all of the time, but this practice can encourage overweight and feed waste. As long as you weigh your rabbits periodically and cut the rations of any overweight animals, feed by whichever method is more convenient for you.

You may use just about any container to feed and water your rabbits, but do not use a container they can chew. We started out with some secondhand heavy crocks for feeders and saucers for waterers. It was not long before we realized that the more expensive self-feeders were a better deal. The crocks occasionally were tipped over and frequently were messed in. Both problems meant lost feed. Self-feeders with a screened bottom are excellent. They hang on the outside of the cage and take up less space. We still use our crocks when we supplement the rabbit feed with cracked corn or oats. If we don't, the rabbits dig through their pellets looking for the "treats," and the pellets end up on the ground.

You may install creep feeders in the cages and feed a high-protein creep feed to the litter from the time they come out of the nest until they are weaned. The feeder is designed so the doe cannot get to the feed. However, a constant supply of regular pellets in the doe's feeder will provide the bunnies with sufficient diet.

Rabbits are by nature nocturnal, so if you feed them only once a day, try to do it at night.

Rabbits love fresh hay and other greens. Buy a well-cured, fine-stemmed hay completely free of mold or mildew. A hayrack fastened to the outside of the cage will cut down on waste. These can be purchased, or you can make them from wire or screening. Never feed greens or other produce to rabbits under six months old. Greens can cause stomach upset and diarrhea, and can kill bunnies.

Oats, cracked corn, and wheat can be fed in small amounts. Feed them in a separate dish to prevent wasting pellets as the rabbits paw through to get to the treats. Corn is a high-energy food, and can help the rabbits through severe winter weather.

Rabbits need salt as a part of their diet. Commercial pellets contain salt, but hang a salt spool (available from feed-supply stores) in the cage to ensure the rabbits a sufficient amount of salt.

Rabbits must have fresh water provided at all times. A doe with a litter needs more water than a dry doe. Rabbits will not eat if they are out of water.

We eventually went to a watering system using a stainless-steel "ballpoint" tube and quart-size plastic bottles. A friend of ours uses this system, too, but in the winter uses crocks because the water doesn't freeze as fast. Whatever watering system you use is fine as long as there is always clean water available for the rabbits, both winter and summer.

Many rabbit breeders give their does concentrated rations to help increase milk production. You may give the doe one tablespoon per day with her regular pellets from the day you put the nesting box in until the bunnies leave the box. We usually wait to start the supplement until the bunnies' milk demands increase, about the fourth or fifth day after birth. If you have spare goat or cow milk, you may feed this in place of the supplement, at a much smaller cost.

Costs

The variations in price of equipment makes it impossible to figure these costs into a projected budget with any accuracy. Most of our cages cost us $10.50 per rabbit; we bought the wire and J-clips and built our own. We have also bought good used cages for as little as $2.50 per cage. The cost of feeders and waterers also varies, depending upon what type you use and whether they are new or used. The cost of equipment can be figured into an annual budget, but at only a small percentage of the total cost; most equipment lasts longer than you will probably want to raise rabbits.

Our cost example uses three rabbits—two does and a buck. We assume they were purchased as mature breeding stock at $20 per animal, and will be used until they are five years old. We assumed a breeding schedule of four successful breedings per year, with each doe raising eight bunnies per litter. We further assumed the slaughter of the litter at eight weeks of age and at a weight of four pounds per bunny.

To determine the amount of feed used per year, we used the feed:meat conversion ratio of four pounds of feed to produce one pound of dressed meat.

CASH COSTS

Breeding Stock	
$60 ÷ 4.5 years	$13.34
Feed	
8 litters of 8 bunnies = 64 bunnies × 4 lb. each = 256 lb. of rabbit dressing out at 60% = 153.6 lb. of meat	
feed: meat conversion ratio yields 614.4 lb. of feed @ $.11 per lb.	$67.59
Total Cash Costs	$80.93
Product Value	
Rabbit Meat	
Sale of half of the rabbits (128 lb.) at $.75 per lb. live weight	$96.00
128 lb. of rabbit dressed for home freezer = 76.8 lb. at $2.60 per pound	$199.68
Total Product Value	$295.68

To end up with seventy-six pounds of rabbit meat in the freezer and $15 above expenses from raising two does and a buck is not an unrealistic goal. The key to success, however, is good management and quality stock. Poor management will cost you in lost litters and feed waste, both of which will adversely affect your budget. Poor stock can provide bunnies that weigh considerably less than four pounds at eight weeks.

MAINTENANCE

Rabbits are not particularly disease-prone, but the best safeguard is good sanitation. If you use hanging, all-wire cages, the job is greatly simplified. The fur that collects on the wire can be burned off easily with a propane torch, or you can vacuum it off.

Washing the cage with a good household disinfectant and a wire brush goes a long way toward keeping a sanitary hutch. A practical system is to clean the cage just before the bunnies leave the nesting box, and again after you have removed the litter from the doe's cage. Always scrub the cage with disinfectant if a rabbit dies.

The care you give the nesting boxes will have a definite effect on your success in raising litters. A doe will not build her nest in a box that is dirty or carries the odor of another doe. Before you put a nesting box in with a doe, make sure you have scrubbed it thoroughly inside and out with a disinfectant, and let it dry in the sun if possible.

If you are using crocks for waterers and feeders, you should rinse them out daily and wash them in dairy or household disinfectant at least once a week. A waterer with a metal sleeve should be washed in a disinfectant once a month to keep it clean.

If you are using a self-feeder, make sure that the dust from the pellets does not build up. Self-feeders can be washed in a disinfectant whenever you clean the cage. Whenever an animal has been sick, wash the waterers and feeders with disinfectant.

Store all feeds in dry containers; never feed wet or moldy food to rabbits. We found a factory that would provide us with metal fifty-five-gallon barrels complete with lids. Once the barrels have been cleaned out, they make great feed bins. The feed is kept dry, and mice and rats cannot get in with the lid closed tight. Not only is it repugnant to find mice and rats in the feed barrels, it is also a health concern, as rodents can carry disease.

SLAUGHTER

Sooner or later you will need to ready some of your rabbits for the freezer. There are two humane ways of making a rabbit unconscious for slaughtering: you can stun the rabbit by a sharp blow at the base of the skull or you can break its neck. We prefer the latter, as there is less chance for error in aim. It takes only about three to five minutes per rabbit to butcher once you develop the skill. The tools necessary are a sharp butcher knife, a filleting knife or hunting knife, pruning shears, and a large pan or bucket of cold water. A hook or nail a couple of feet over your head and a piece of twine are needed to hang the rabbit. You also need a container under the rabbit in which to put waste.

Reach into the cage with your stronger hand and remove the rabbit by the nape of its neck. Then get a good grip over the hind legs with your weaker hand. Next place your hand on the rabbit's head, your thumb behind the ears on top of the neck bone. Your fingers will be under the jaw. Now simultaneously and rapidly extend the rabbit across in front of you. With one sharp motion, press down hard on the thumb and up on the fingers on the jaw, while stretching the rabbit. This breaks its neck.

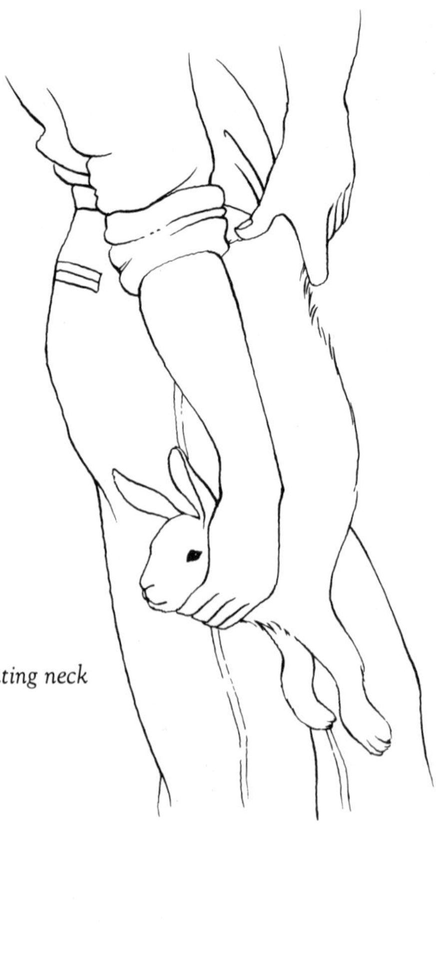

Dislocating neck

Immediately hang the rabbit up by one leg, looping the twine above the hock. Grasp the ears and with long strokes with the butcher knife, cut away from you, removing the head just behind the ears. Grab the front legs so the rabbit bleeds away from you. When the bleeding slows to a trickle, remove the front legs at the paw joint and the one free leg at the hock with the pruning shears.

DRESSING OUT

Remove the skin with a filleting or hunting knife. Cut the skin completely around the leg on the rabbit side of the twine from which it is suspended. Then slit down to the tail. Now pull the skin away from the tail and from the loose rear leg. Last pull the skin downward off the neck and front legs. It will come off wrong side out. Skin will still be on the tail and genitals. Hold the tail and genitals in one hand and remove them with a knife by cutting on both sides, forming a V to remove them. Then cut the underside of the pelvic bone.

To remove the guts, slit the belly down from the pelvic bone to the rib cage. The knife is inserted at the pelvis and the blade is run close

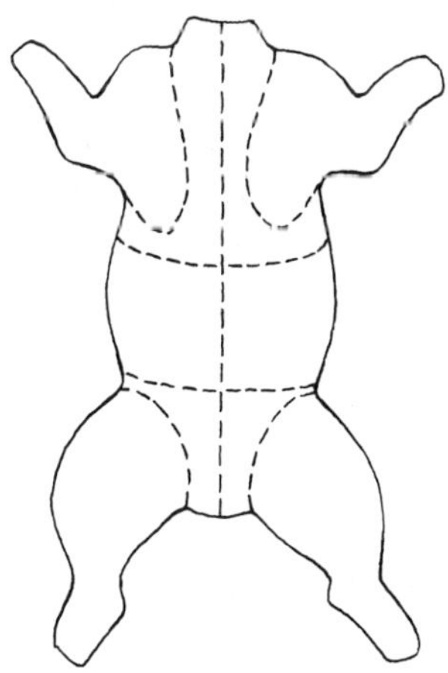

Guidemarks for butchering a rabbit

to the skin so as not to cut the intestines. It will not take much pressure to slit the belly. Reach palm down into the cavity; remove the guts, being careful not to tear the intestines or gallbladder. Separate the liver from the mass; put it in the cold water and discard the mass. Then insert the knife point into the oval hole in the diaphragm by the spine and split the ribs all the way down, including the skin under the neck. Remove the lungs, heart, and esophagus. With your finger, remove the two pieces of diaphragm which are still on the carcass and the two strips of back fat.

Cut the carcass loose from the hook, and put it in the cold water. To ensure a minimum of water absorption and to maintain the pearly whiteness of the meat, do now allow the meat to sit in the cold water for longer than twenty minutes.

Cutting up the carcass is also easy. You should have seven pieces, not counting the liver. Remove the front legs by cutting between the shoulder blade and ribs and around the shoulder joint and each leg. Now twist the legs, one at a time, out of the sockets. Remove the rib section by cutting to the backbone just behind the ribs on each side. Then cut in from the flanks where the flanks and loin meet at the backbone on each side. Last, twist the ribs from the loin and the loin from the rump. You will be butchering at least eight to twelve rabbits at a time; package all the livers together for a real delicacy.

Rabbit meat is both delicious and good for you. The U.S. Department of Agriculture provides the following analysis:

CONTENT PER LB.

Meat	*Protein*	*Fat*	*Water*	*Calories*
Rabbit	20.8%	10.2%	27.9%	795
Chicken	20.0%	11.0%	67.6%	810
Beef	16.3%	28.0%	55.0%	1,440
Lamb (med. fat)	15.7%	27.7%	55.8%	1,420
Pork (med. fat)	11.9%	45.0%	40.0%	2,059

With the low moisture content, there is very little shrinkage when the meat is cooked. When you cook a pound of rabbit, you put almost a pound of meat on the table.

Fryer-broiler-weight rabbit is two to two and a quarter pounds dressed out of four to five pounds live weight. Any rabbit over five pounds live weight is best used as a stewer, in a casserole or salad, or as a roaster. You can use any chicken recipe for rabbit, and there are a number of rabbit cookbooks. One of the best cookbooks is the *Domestic Rabbit Cookbook*, published by the American Rabbit Breeders Association. Our family cannot decide whether our favorite rabbit recipe is oven-fried or barbecued.

Medical Concerns

Rabbits are not prone to disease, so if you are providing good nutrition, fresh water, and sanitary conditions, you should not have major problems. Good breeding stock also helps prevent medical problems. A few diseases you might encounter are described here. Any medications should be given in the dosage recommended on the medication label. A normal temperature for a rabbit is 102.5° F (39° C).

Mucoid enteritis is the leading cause of death among young rabbits, ages four to eight weeks. The primary symptom is diarrhea. The bunny will also have a hunched appearance, squinted eyes, and drooping ears. There is no real cure for this disease. If a rabbit develops mucoid enteritis, you can try the soluble powder used to combat blackhead disease in turkeys; this is mixed with their water from the time the bunnies leave the nesting box until they are weaned. However, do not slaughter a rabbit for meat that has been treated with this medication until at least one month has passed from the last dosage.

Pneumonia is the single largest cause of death among mature rabbits. The symptoms are depression, labored breathing, nasal discharge, and slightly elevated body temperature. To treat, give an intramuscular injection of a combination of penicillin and streptomycin. Inject the solution under the fold of skin on the back of the neck or into the muscle of the hind leg. Feel for the muscle to make sure you do not hit the bone. Eliminating drafts and excessive dampness is crucial in avoiding pneumonia in your rabbitry.

Sore hocks are developed by the rabbit's stamping its hind legs. It was cute when Thumper did it in *Bambi*, but a frequent thumper in the rabbitry is no pleasure. The tendency toward sore hocks is hereditary, and if the problem is severe, the animal should be culled. To treat the problem, remove any fur from the area, wash with soap and water, and apply a topical antibiotic until the condition heals.

Malocclusion or wolf teeth is a condition in which the lower jaw is shorter or longer than the upper jaw; as a result, the teeth keep growing instead of being ground down naturally by chewing and gnawing. You can cut the teeth back with clippers, but it is a temporary cure. As this is an inherited tendency, you would be wise to cull the offending parent.

Ear mites cause ear canker and may eventually prove fatal. If you have a rabbit that shakes its head a lot or rubs its ears, check for crusty scales in the ears. To treat, swab the entire inner ear with a cotton applicator stick soaked in mineral oil, and then massage the ear to work the oil down. Remove all the crust and debris as it softens with the oil. Ear mites are highly contagious, so treat all your rabbits and dispose of the equipment you used to clean the ears. Do this twice a day and you should see improvement within a day or two. If your rabbits do not respond to the mineral oil, there are ear-mite insecticides available.

Conjunctivitis or weepy eye is an infection of the membrane that covers the surface of the eye and the inner part of the eyelids. There usually is a thick white discharge from the eyes. This can be caused by dust and other irritants, so check your bedding conditions. Also check bunnies' eyes when they are supposed to open (about ten days old) to make sure they do not have a problem. To treat, swab eyes with an over-the-counter eye wash, then apply an antibiotic ophthalmic ointment. Continue treatment until the condition clears up.

Coccidiosis or spotted liver is caused by a parasite that invades the intestine or liver. The first symptom of this disease is the presence of discolored spots on the liver. Always examine the livers when you are slaughtering rabbits. If left untreated, the infestation can increase until it is fatal. To treat, add sulfaquinoxaline to the drinking water. Wire cages help avoid this problem.

Heat exhaustion can be a killer. If your rabbits are housed outside, make sure they have shade. On extremely hot days, place wet feed sacks on the hutch floor. Glass jars or crocks filled with ice or ice water will help.

Occasionally you may find a case of cannibalism with a doe and her litter. This may be caused by inadequate or insufficient diet. Generally, however, it is due to the doe's being disturbed after kindling. This disturbance can be a horn honking, children playing, dogs, cats, mice, rats, or any number of things.

Raising rabbits is fun, money-saving, and provides delicious and healthful meat. However, success is not all that easy. If you get the does bred, and if they raise good-sized litters, you have a productive experience. Things can and do go wrong, however. You will probably waste some time thinking does are bred when they are not, and you may lose an occasional litter. If the loss of production is your fault, improve your system; if it is the doe's fault and she repeats the problem a second time, cull her.

Never become friends with any rabbits but breeding stock. All our breeding stock are named and babied, but bunnies going into the freezer, never. Avoid letting your children play with the bunnies or the breeders. Injured breeding stock is no good to you, and the bunnies are too endearing to play with and then eat. To avoid these problems, our Katie has a pet Dutch rabbit that she plays with any time rabbits seem irresistible to her.

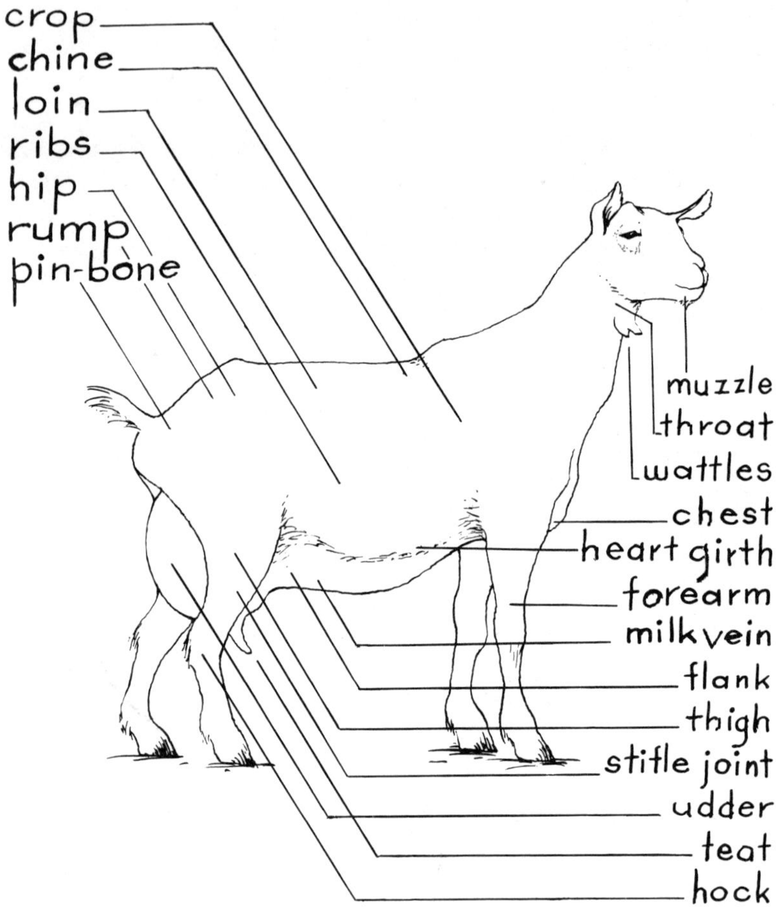
crop
chine
loin
ribs
hip
rump
pin-bone
muzzle
throat
wattles
chest
heart girth
forearm
milkvein
flank
thigh
stifle joint
udder
teat
hock

3

GOATS

DICK BEGAN talking about the advantages of owning a goat in May of our first year. When he saw some registered milking goats advertised in our local newspaper, he suggested seeing the goats over the Memorial Day weekend.

Our visit with Mrs. Hughes and her eight goats and six kids was great fun. She proved to be a delightful, knowledgeable woman who patiently reassured us that goats do not eat tin cans, clothes off clotheslines, or any other bizzare items. In fact, goats are rather fussy eaters. They get their "eat anything" reputation because they consume such items as brush and poison ivy.

Another myth was exploded when Mrs. Hughes assured us that female and baby goats are not odoriferous, although the buck goat deserves its reputation of having a strong smell.

Mrs. Hughes taught us a few terms that you should know before you buy a goat. The female of the species is a doe, not a nanny, and the male is a buck, not a billy; a milking doe is properly called a milch goat. The average weight of a doe is 140 pounds, that of a buck 190 pounds. A castrated buck is called a wether. Baby goats up to six months old are called kids, and growing kids up to a year are doelings, or bucklings. A doe giving birth is said to be kidding or freshening; a doe that has had her first kidding is called a first freshener.

Mrs. Hughes wanted to sell a Nubian doe named Lola; she had two does, both of whom wanted to be boss of the herd. Lola's managerial skills have since proven to be a dominant factor in her personality. Dick, thoroughly taken with Lola, paid Mrs. Hughes $150 and loaded the goat into our station wagon. It was our first experience with a goat in a car, and Lola loved it. She bleated enthusiastically at everyone she saw, causing several near-accidents.

Back at the farm, Tequilla, our half-Arabian mare, had gorged herself on the summer's first cuttings of grass and developed colic. We got her up and walked her awhile and then put her in the barn with some hay.

Lola resented the attention Tequilla got from us; she seemed convinced that her arrival called for considerably more fanfare. She felt amends had been made when our son Rome slept in the barn that night. He slept there to keep an eye on Tequilla, but you could not convince Lola of that. Since that night, Lola has always had a special regard for Rome and is happiest when working with him.

Her favorite work project was building a rail fence along the front of our property. Rome tethered her out every morning that summer, near where he was working. We figured the fence construction would take him a couple of days, but with Lola's help it took him two weeks. Goats are curious creatures, and Rome did not dig one post hole or put up a single rail without Lola's verbal and physical comment on each movement.

We tethered her out so she could graze on the grass. We had her about six weeks when I read that goats do not like to eat grass; as is my custom when our animals are not performing according to the book, I read the article aloud to her. I do not think it made much of an impression on her, as she just kept eating the grass.

Goats and people go back together a long way. The earliest records indicate domesticated goats in Iranian villages during the fifth millennium B.C. The common goat, *Capra hircus*, belongs to the same family (Bovidae) as the sheep. Goats are gregarious by nature and do not like to be kept by themselves. If one goat is all you need or want, house it with sheep or a horse to provide companionship. Housing and equipment for sheep and goats are of the same design and size, which makes them easy to house together.

BREEDS

The five main breeds of dairy goats in the United States are the Nubian, Toggenburg, Saanen, French Alpine, and La Mancha. Other breeds found in this country include the Rock Alpine, Swiss Alpine, Angora, and African Pygmy.

The most popular purebred goat is the Nubian, which traces its ancestry to Egypt. Nubians are the easiest breed to recognize: they have pronounced Roman noses and lovely, long, droopy ears, giving them an almost pharaohlike appearance. Nubians can be almost any solid color, parti-colored, or spotted.

The average Nubian produces less milk than the average goat of any other breed, but the butterfat content is higher—about 5 percent, compared to about 3.5 percent for other breeds. For this reason, many people think a Nubian's milk tastes best.

Every Nubian I have ever known has had a certain arrogance and regality mixed with a loving and curious disposition. You may hear it said that Nubians are the noisiest of the breeds; I would not say they are exactly noisy, but they are great talkers.

Toggenburgs are the oldest registered breed in the world; a herd book from the Toggenburg Valley in Switzerland dates back to the seventeenth century. Toggs have erect ears, slightly concave faces, and are always some shade of brown with a light or white stripe down each side of the face and white on either side of the tail, on the rump, and on the insides of the legs. Toggenburgs have a sweet and gentle disposition. They are just as curious as Nubians, but much less aggressive.

Saanens, originally from the Saanen Valley of Switzerland, resemble Toggenburgs, but are pure white or cream-colored. They, too, have erect ears and concave faces. Saanens are said to give the most milk. Some purebreds have produced more than five thousand pounds of milk a year—a lot of cheese wheels!

The French Alps contributed the French Alpine breed. Its coloring ranges from white to black, often with several colors and shades on the same goat. The color patterns of this breed are classified as follows: Cou Blanc (white neck), Cou Noir (black neck), Chamoisee (color and markings similar to a chamois), and Sundgau (black with white underbody or white Togg markings). French Alpines are also sweet and gentle; they seem to me to be less assertive than either the Togg or Saanen.

The American La Mancha is easily identifiable because it is almost earless. The breed was developed in this country from a short-eared Spanish breed crossed with other purebreds. The La Mancha kids look like adorable little teddy bears; the breed is known for its gentle disposition.

The African Pygmy goat recently had a herd book established by the American Dairy Goat Association, which means it is a registrable breed. The Pygmies are twenty-one inches tall or less at the shoulders and vary in color. Despite their size, they are excellent milkers. They are also more apt to have triplets or quadruplets than other breeds.

The Angora is primarily known as a wool producer. Its body is covered with soft hair four to six inches long, which is spun into mohair.

Classification

Once you have decided which breed of goat you prefer, you need to decide whether you want a registered animal. Goats are classified as registered purebred, unregistered purebred, American, registered grade, and grade. The registered purebred is an animal that has pedigree papers showing its ancestry and accompanying registration papers showing that it is entered in the herd book of a registry association. The American Goat Society and the American Dairy Goat Association are the two registry associations that maintain herd books on dairy goats in the United States. An unregistered purebred is just what its name states. It is difficult to prove that an unregistered purebred is indeed purebred.

The classification American means that the animal is $\frac{15}{16}$ pure; in other words, one of its sixteen ancestors was a registered grade. A grade is any animal without a pedigree; it may be purebred, but this cannot be proven. Most grades are a mixture of two or more breeds. A registered grade has met certain requirements set up by the American Dairy Goat Association and is recorded as a grade of whichever breed it most closely resembles.

Two other terms you may hear while searching for your goat are Star Milker and Advanced Registry. The Star Milker certificate is earned by a doe on the basis of an official one-day test of milk production, butterfat content, and length of lactation. For a Nubian to qualify, she must produce at least eight pounds of high-butterfat milk within the given twenty-four-hour period; a Togg or Saanen must produce at least ten pounds of milk in that time. Goat's milk is measured by weight. It weighs about one pound per pint; a gallon milker produces about eight pounds of milk a day.

The Advanced Registry certificate is based on an official testing of the doe's full lactation period (305 days). A doe must produce at least fifteen hundred pounds of milk and equivalent butterfat, depending on her age at freshening time, to qualify for the Advanced Registry certificate.

You can be sure that a Star Milker or an Advanced Registry doe is a good milker. Sometimes you will see two-star and three-star milkers. This means that not only is the doe herself a star milker, but also that her dam (mother) and grandam have earned their stars.

CHOOSING YOUR GOAT

Whatever breed or classification you select, make sure to check the animal's overall appearance. The more goats you have seen, the easier it is to assess an individual goat; attending goat shows is an effective way of developing a keen eye. A goat should have a straight back and a long rump not too steeply slanted toward the tail. The hips should be slightly higher than the shoulders and neck length neither too long nor too short in proportion to the rest of the body. Dairy goats should be thin but not scrawny.

Look for a broad chest, a sign of a strong respiratory system. The nostrils should be large and well distended. The legs should be straight and the goat should stand squarely on its hoofs. Check the hoofs carefully; a goat from a good home will have nicely trimmed hoofs. Do not buy a goat with any sign of lameness.

A goat's approximate age can be determined by its teeth, using the same standards applied to sheep (see Chapter 5). Adult goats by age three to four should have eight bottom incisors and twenty-four molars; they do not have any top incisor teeth.

Gender

If you want a milch doe, pay particular attention to the udder. Biggest is not best; an udder may have become enlarged because of excessive muscle tissue, and the risk of injury to a pendulous udder is great. Avoid goats with signs of mastitis (inflammation of the udder), indicated by a hot, swollen, hard, or lumpy udder. Udders should be pliable, soft, and fairly symmetrical.

The teats (nipples) should be uniformly placed, large enough to be milked easily, and point slightly forward. There should be only two teats with single orifices, and they should be approximately the same size. Both very large and very small teats are undesirable.

If you are planning on using your goat for your family's milk supply, by all means ask to taste the goat's milk before you buy her. The milk should be pure white, sweet, and tasty; if it is not, you do not want that goat. Good goat's milk tastes similar to cows' milk: sweet and refreshing (in fact, some people prefer goat's milk).

A milch goat generally gives milk for ten months a year if you have her bred annually. You can continue to milk during the first three months of gestation, then dry the doe off two months before kidding. When buying a doe for the family dairy supply, ask when she freshened, so you know how many months of milking you have before you need to have her bred again. Most goats will continue to give milk beyond ten months even if they are not rebred, but the further from the last freshening, the less milk you will get. One family I know milked its doe for two and a half years before having her rebred and were still getting a quart a day, but I suspect this is quite unusual and not a very good policy.

Carefully consider whether you should own your own buck. Our does would probably be delighted to have a buck in residence, but we decided against one. Keep in mind that male goats do have a strong smell. The aroma may excite your does, but it will undoubtedly turn off your friends and neighbors. Bucks also are prone to spraying their urine and semen on their beards and forelegs, which does not help them win any nongoat friends. It only costs from $15 to $25 to have your doe bred to a registered buck, far cheaper than keeping your own.

A 200-pound powerful buck, especially during mating season, can be difficult to handle and certainly requires more elaborate housing and fencing than a doe. The buck must be kept apart from the does at all times, except for the minute it takes to have the buck breed the doe. Housing a buck with milch does is apt to ruin the milk's flavor. I think it is safe to say that buck ownership is not for the average homesteader.

Age

The age of your prospective goat is also something to consider. The normal lifespan of a goat is eight to twelve years; the average goat will be productive for up to ten years, and goats have been known to kid at twelve and thirteen with no problems. For maximum production and longevity, it is probably best to start with a two- or three-year-old doe that has recently freshened. A doe's best milk-producing years are generally from age three to six.

Horns

Some goats are naturally polled (hornless), others have been dehorned, and some have horns. The absence or presence of horns has nothing to do with either the gender or breed of the goat. Horns on goats are a bother; they cause serious injury to other livestock and to the goat itself, should the horns get caught in fencing or hayracks. A gentle goat can unintentionally injure a child or an adult by a misguided swing of the head. If your goat has horns, they can be removed; see "Dehorning and Disbudding," page 77.

HOUSING AND EQUIPMENT

Adequate housing for goats is relatively simple to provide. A partitioned section of a barn, tool shed, garage, or chicken coop is a sufficient shelter. The main criteria are that the structure be dry, draft-free, and receive adequate light.

The amount of space per animal depends on how much time the animals will spend in the enclosure. If you have only a small outside area allotted for the animals, you should provide sixteen to twenty square feet per animal; twelve to sixteen square feet per animal is sufficient if there is ample outdoor space for exercising. In determining how many goats you can house in your structure, do not forget the kidding season.

Goats can be housed with or without individual stalls. A loose housing system allows the goats to roam freely and cuddle up together on cold nights. Use deep litter for bedding and add more bedding material when needed. This eliminates a daily manure detail, for with this system the shelter only needs to be cleaned out once or twice a year. The depth of the bedding material also provides insulation during cold weather.

The flooring can be dirt, concrete, or wood. If your floors are wood, the deep-litter system helps keep the wood dry and retard rotting. If you have concrete flooring, use a platform or sufficient bedding to keep the goats dry and comfortable.

Use the most absorbent bedding you can find at a reasonable price. Peat moss is best by a considerable margin, but it is expensive; chopped straw, wood chips, or sawdust are more economical.

Ideal housing includes running water and electricity; of the two, electricity is more important. During the winter, it is still dark at milking time, and electric lighting is more than a luxury, and is certainly safer and more convenient than a lantern.

An exercise yard contributes greatly to a goat's health and happiness. It is best if the yard is so situated that the goats can go directly to the yard from their quarters by themselves. This arrangement frees you from having to run the goats inside when it rains: goats do not like to get wet.

The fence for the exercise yard may be of boards, woven wire, or barbed wire. Woven wire should be strong enough to withstand the goats' weight if they jump or stand against it. You can keep the goats off the fence by stringing a strand of electric wire along it $2\frac{1}{2}$ feet from the ground.

For fun, you can supply a small mountain for the goats; just a hill of dirt or a barrel will do. The time you spend supplying your goats with "toys" will be well rewarded, especially when the kids arrive. Goats enjoy playing with homemade seesaws, old tires, tree stumps, and each other.

Most goatherds feed their does while milking them and have the milking stand equipped with a feeding pail. Doelings and dry does should also be fed while on the stand to ensure consistency and cooperation.

Hay should be fed from a manger, hayrack, or a combination hay and grain rack to minimize waste. We successfully use an old porch railing for the slatted side of our hay and grain rack. Keyhole shapes cut into boards are a highly satisfactory design for a hayrack.

Watering equipment must be sufficiently large and easy to clean. Automatic waterers are ideal, and they are not very expensive if you have running water in your goat quarters. If not, you may use large galvanized tubs.

Goats will not drink water from a dirty container. Rinse the water container every day before providing fresh water. Once a week, fill the tub with water and bleach, let it sit for fifteen minutes, then scrub with a brush. Remember that water is needed for milk production, and clean water tubs mean more milk in your pail.

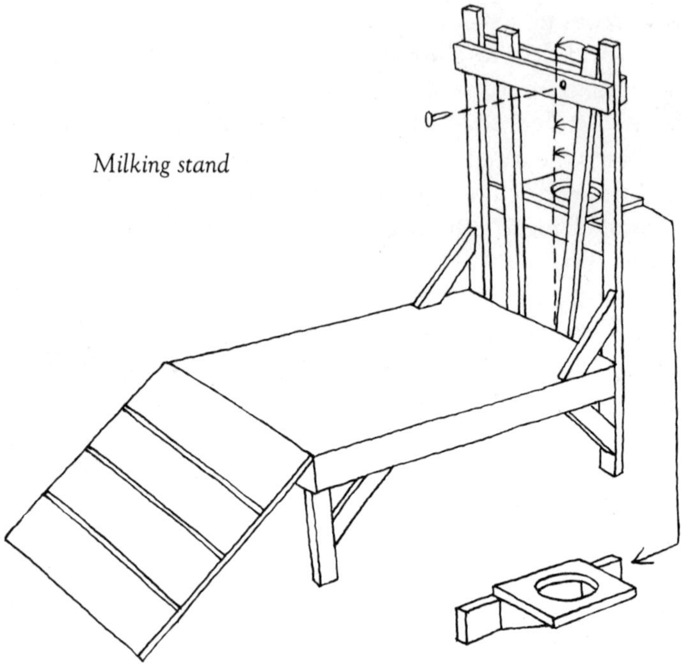

Milking stand

MILKING

The night we brought Lola home, we did not have a milking stand ready and she had become accustomed to one. That first milking was not a particularly satisfactory experience for any of us. We ended up with two of us holding her against the wall and a third person milking and trying to keep her from kicking over the milk bucket. Build or buy a milking stand before getting a milch goat; it aids in maintaining sanitary conditions and simplifies the whole matter.

A milking stand consists of a ramp, bench, stanchion (vertical post), feeding tray, and seat. If your milking stand is not against a wall, construct a railing to ensure that the goat does not fall off the side opposite the person milking. There is no correct side to milk a goat, but they are creatures of habit; once they get used to being milked on one side, they seem to prefer it.

GOATS

Goats should be milked every twelve hours for maximum production. Whatever hours you can best fit into your schedule are fine, as long as you are regular. We milk our goats at 7 A.M. and 7 P.M. If you have more than one goat to milk, establish a sequential order for milking and stick to it.

When it is time to milk, stanchion your doe on the milking stand with her grain ration. Using a paper towel, wash her udder with warm water and an udder-washing solution or soap. Dry her udder and your hands. This is important not only for sanitation, but the massaging motion of washing and the warm water assist in bringing the milk down. Wait about one minute after washing the udder before milking.

To milk, sit beside the goat, facing to her rear. Place the pail under the udder and take each teat between your thumb and index finger, encircling it near the base of the udder. (Be sure not to grasp the udder, because you could cause tissue damage if you milk too high.) Close your thumb and finger together and hold firmly; this prevents the milk in the teat from going back up into the udder. Then firmly but gently squeeze the teat first with your second finger, then with your third finger, and finally with the little finger.

Milking procedure

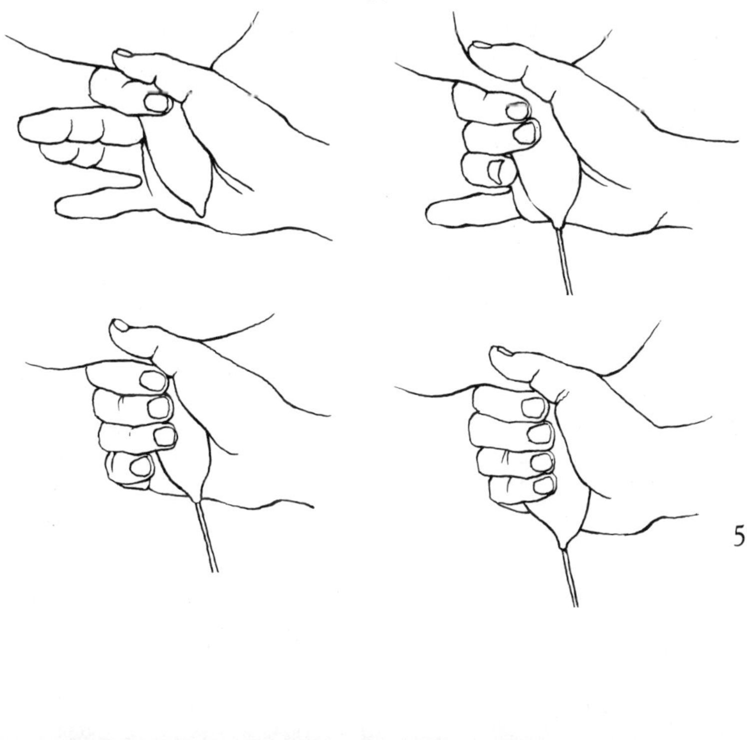

With each squeeze, you force the milk down the teat and, with luck, into the pail. Release the teat, and while it is filling up with more milk, repeat the process with your other hand on the other teat, alternating until you have finished milking. An udder is not like a bag filled with milk, but rather like a sponge.

Do not squeeze too hard and keep your motions smooth. There is

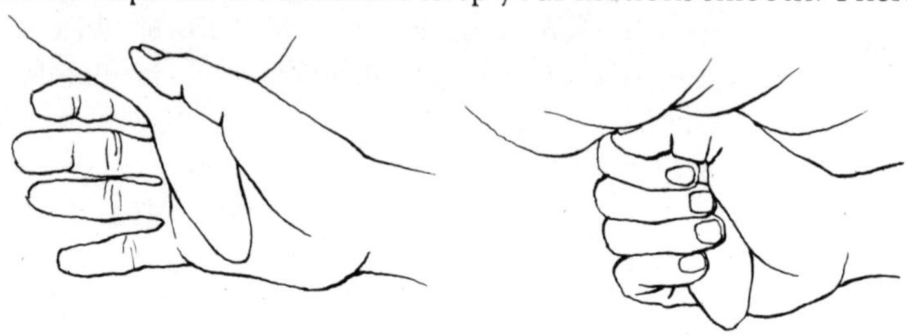

a rhythm to it and you will quickly catch on; however, there is a definite advantage to having somebody who knows how to milk teach you. Do not wait until your goat first freshens to learn; at that point she has had enough excitement without having to put up with the nervous fumblings of a novice milker.

When you stop getting milk, massage the udder or butt it with your hand like a kid would, and then milk again to see if more milk has come down. The final bit of milk should be stripped (milked) out by running the thumb and index finger down the length of the teat. Leaving a little milk in the teats may contribute to mastitis, so do not forget this last step.

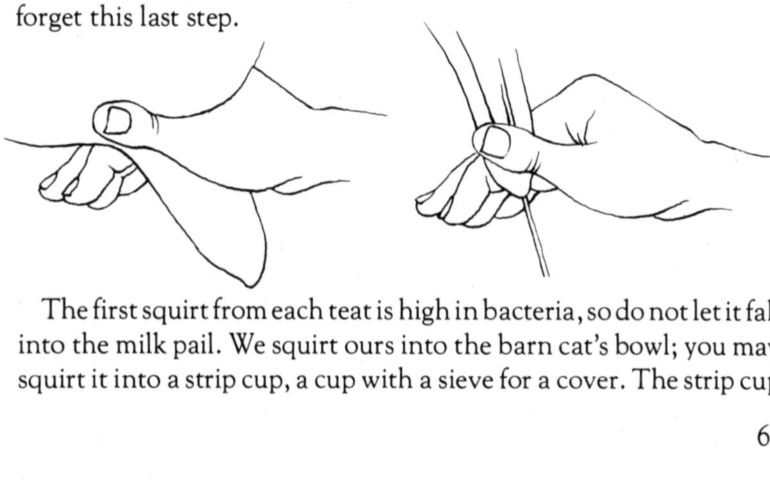

The first squirt from each teat is high in bacteria, so do not let it fall into the milk pail. We squirt ours into the barn cat's bowl; you may squirt it into a strip cup, a cup with a sieve for a cover. The strip cup helps detect lumpy or stringy milk, a sign of mastitis.

You may use any pail to milk into, but the four-quart seamless stainless-steel pail designed for goats is well worth the money. The advantages of the professional pail are that it is the right size, has a partial cover, and is easy to clean.

Cleaning the milking equipment thoroughly is very important. If the equipment is not handled properly, the milk will spoil quickly and/or taste peculiar. Rinse all equipment that has contained milk immediately with cold water, then use hot soapy water and a brush to wash it. Feed stores stock dairy-cleaning agents, such as iodine compounds; wash your equipment daily or as directed.

Care of the Milk

After milking, strain the milk to eliminate any foreign objects, such as hair. We pour the milk through a filter in a sieve that rests in a funnel set into a half-gallon glass jar. Professional strainers with disposable filter discs are available from dairy catalogs or feed stores.

Cool the milk as soon after milking as possible. Your goal is to chill the milk to 45°F (7.2°C) within an hour after milking. Prechilled milk containers placed in the freezer section of a refrigerator for an hour will accomplish the goal. Remove the milk before it freezes.

To pasteurize the milk, heat it rapidly to a temperature of 165°F (74°C), stirring constantly. Use a stainless-steel pan and stirring implement, as other metals may impart an undesired flavor to the milk. Keep the milk at 165°F for twenty seconds, then immediately place the pan into a large bowl of cold water. Stir constantly to reduce the temperature quickly to 60°F (15°C). Store the milk in a covered container in the refrigerator. Pasteurization is not necessary if the milk is for home consumption and therefore to be used in a relatively short period of time. If you are making cheese, pasteurize the milk.

Goat's milk tastes much like whole cow's milk and is nearly always pure white in color. It is healthful and nutritious, and more easily digested than cow's milk because of its small fat globules and soft curd. In addition, goat's milk is rich in antibodies and has a much lower bacterial count, when freshly drawn, than cow's milk.

Goat's milk is particularly sought after for infants with a fat intolerance or acidosis. Some people suffering from ulcers, dyspepsia, allergies, insomnia, and morning sickness have benefitted from drinking goat's milk. It is certainly not a panacea, but it is easier on many people's digestive systems than cow's milk. Far more people in the world drink and use goat's milk than cow's milk. It can be used to make butter and both hard and soft cheese, Feta cheese being a well-known example.

Even though goat's milk is said to be naturally homogenized, you can make butter from it. Separate the cream from the milk with a cream separator, or proceed as follows: cool the milk and then leave it undisturbed for eighteen to twenty-four hours (milk kept in one large container will separate more slowly than milk in several small containers). Set the milk on the stove without disturbing the top cream. Heat the milk very slowly until a light wrinkly skin forms on the surface. Place the milk back in the refrigerator, being careful not to disturb the top cream, for another eighteen hours, then skim off the cream.

To make butter, beat the cream by hand or use a churn or electric beater; as you beat it, most of the buttermilk will separate from the butter clumps. Once the butter clumps are well formed, pour off and save the buttermilk. Press the butter repeatedly against the sides of the bowl with a wooden spoon to force out the buttermilk. Cover the butter with cold water and continue to press out buttermilk. Keep rinsing until the water is clear; buttermilk left in the butter will shorten the length of time it stays sweet. Butter made from goat's milk is a little softer and smoother than butter from cow's milk. It is pure white, so you may add a few drops of yellow food coloring if you want to have the familiar butter color.

Recipes for both hard and soft cheese are included in Chapter 11, "Homesteading Reference Guide."

BREEDING

The primary breeding season for goats is from September or October through December, but we have had does bred as late as

April 1. Nubians in particular tend to have a longer breeding period. The gestation period is 145 to 155 days. Does usually have twins, although one or three kids at a kidding is not uncommon.

Doelings may be bred when they weigh eighty pounds or more and are at least eight or nine months old. Breeding them too early is bad for their health and milk production. Young bucks are capable of breeding at three or four months, which is important to remember if you have an uncastrated buck kid.

During the breeding season, does are fertile for about twenty-four to thirty-six hours at more or less regular intervals of about three weeks until they are bred. A doe in heat may wag her tail violently, bleat nervously, and have a slightly swollen vulva, which may be accompanied by a discharge. Some does, however, give very little indication of being in heat.

A buck rag is helpful in knowing when to breed your doe. This is a piece of cloth which has been rubbed over a buck and stored in a closed container. Show it to a doe you suspect is in heat and if she starts rubbing her body against anything handy and bleating after smelling the cloth, take her to him. The doe only needs to be serviced once, unless she comes into heat again in three weeks, in which case she must be rebred.

Try to choose a buck that comes from a line with good milkers as well as watching for good health and body conformation. If you have a color preference for a kid, the general rules of genetics should be considered; for example, dark coats are dominant over light ones.

To improve the chances of your doe having a multiple kidding, you may try flushing—feeding your doe extra grain for two to three weeks before and after you have her bred. The extra feed helps promote the release of more eggs from the ovaries and, after breeding, helps the development of the ovaries in any female fetus.

Methods

There are four breeding methods that can be used. One is inbreeding, the breeding with close relatives (father-daughter, sister-brother). Inbreeding is not recommended for goats because existing faults will only be accentuated and perpetuated in the offspring.

Linebreeding is the mating of less closely related animals, such as aunt-nephew, with the idea of retaining a close genetic relation to some excellent ancestor. Outcrossing is the mating of unrelated animals within the same breed. Outcrossing can be highly successful, especially if you know that a line consistently produces some characteristic that you are trying to breed into your line. Crossbreeding is the mating of two different breeds, for example a Toggenburg with a Nubian. Linebreeding, outcrossing, and crossbreeding are all good breeding methods for goats.

KIDDING

The doe's gestation period may be divided into two parts: the first three months and the last two months. Each period requires specific feeding practices. If the doe is a first freshener, feed 1 pound of grain daily, plus 2 to 3 pounds of hay, for the first three months of gestation. Milch does should be fed according to their milk production—$\frac{1}{2}$ pound of grain per each pound of milk—until they are dried off. During the last two months of gestation feed $1\frac{1}{2}$ pounds of grain, plus 1 to $1\frac{1}{2}$ pounds of hay, twice a day until about a week before kidding. At that time cut back to $\frac{3}{4}$ to 1 pound twice a day and increase the hay to 3 to $3\frac{1}{2}$ pounds. Make increases and decreases of grain rations gradually.

Provide the doe with an individual stall, or use the kid's pen for kidding. You may put her in it on the 145th day or wait until she is almost due to kid. The stall should be as clean as possible, with plenty of fresh clean bedding. When kidding time is close, offer the doe water when you are checking her and then remove the pail. She will not be very thirsty at this time and the water pail could be a danger should a kid be dropped in it.

If you are going to bottle feed the kids, set aside a separate pen for them; this may be a lambing pen or any fenced-off area. The kids will be separated until they are weaned at three to four months, so the pen should be about four feet by four feet to provide sufficient room, and have sides at least four feet high to keep the kids from jumping out. Use lots of clean hay or straw for bedding.

By the 145th day after breeding, check your doe frequently. Get your kidding equipment ready; you will need clean towels, iodine or boric acid, a small cup or wad of cotton, a pessary (medical vaginal suppository), a lubricant such as mineral oil, a germicidal soap, and a pair of sterile surgical gloves.

The doe's udder will begin to fill with milk several weeks before kidding; this is called "making bag." The udder should remain flexible, and if it becomes hard, shiny, and loses its flexibility, milk out a small quantity.

Several days before kidding, the doe will probably be nervous and restless, her flank on either side of her tail will appear hollow, and mucus may be discharged vaginally. When kidding starts, thick mucus with strings of blood is passed.

The day is at hand when the doe becomes very restless. She will probably do lots of standing up and lying down, pawing the bedding, and bleating, and she may refuse her grain that day. These symptoms may not appear, however, as some does are calm and eat right up to kidding time.

The hour is at hand when thick mucus starts passing and the doe goes into contractions. The birth canal widens and the amniotic sac becomes visible. Sometimes the sac remains intact, but it can break at this point or early in labor. The front hoofs of the kid usually appear first, with the head between them or resting on them. The doe will then rest briefly. If the sac is intact, break it. If you break it or if it is already broken, the kid will probably cough or sneeze to clear its breathing apparatus. If not, try to remove the mucus around the kid's nose and mouth with a soft, clean towel. One or two more contractions and the kid emerges. If your doe prefers standing through her delivery, you can catch the kid on its way to the floor.

Sometimes the kid comes out back end first, also a normal position. You will recognize this position because the kid's toes point upward instead of downward as they come down the birth canal. Watch for the breaking of the umbilical cord; once it is broken, the kid starts to breathe and should be out of its mother's body as quickly as possible. When the kid's shoulders appear, pull firmly downward as the doe contracts. Do not jerk the kid; provide a nice, steady, firm pressure downward, not straight out. Work with the doe. Once the kid is out, clean its nose and mouth as described above. If the kid is not breathing properly, hold it upside down by its hind legs and slap its side firmly with an open hand; this forces the rest of the mucus and fluid out.

Give the kid back to the doe and she will clean it off. When she is done, disinfect the kid's navel by pouring iodine into a small cup, placing the cup over the navel and tipping the kid upside down. Then put the kid in the kidding pen.

If the temperature is near freezing or below, you can place a blanket in the pen and use a heat lamp for a short while, but do not let the kid get too warm. Remember, it will have to adjust to the barn temperature and you do not want it to get accustomed to a balmy existence in midwinter. If the kid shows signs of freezing (acute shivering and tissue discoloration), place it in a pail of warm water up to its nose for a few minutes. Pay particular attention to the large ears of a Nubian kid. When the kid revives, dry it off completely, wrap it in a blanket and put it in the pen.

If twins have been kidded, the doe is probably at work on number two by the time you finish with the first kid. Repeat the whole process on the second kid. If a half hour passes from the last birth and the doe seems comfortable, it is safe to assume that kidding is over. See that she gets fresh hay to eat and a bucket of warm water to drink. You can also give her a bowl of bran moistened with warm water as a special treat. Remove any soiled bedding and put down lots of fresh bedding in the kidding pen.

Once the doe has passed the placenta you can either bury it or let her eat it. She may pass it quickly or take hours, but if she has not passed it after twelve hours call a veterinarian. Never pull on the placenta to help the doe pass it; milking the doe or having the kids nurse after you are sure kidding is over helps the uterus contract and expel the placenta. If you are not going to have the kids nurse, wash the udder with warm, soapy water and dry it before milking.

After you have cleaned the mucus out of the kid's nose and mouth, check the umbilical cord. It will probably have broken during kidding; if not, squeeze the cord with your thumb and forefinger for a few seconds to stop the blood flow and the cord will break. Some authorities recommend that you cut the cord with scissors several inches from the kid's stomach, then tie a string on the cord two or three inches from the navel. We have found that this procedure is not necessary.

The kids are usually not hungry for a couple of hours after birth. The first milk after kidding is called colostrum; it is a thick, yellow liquid loaded with antibodies and vitamins. The doe produces colostrum for the first three or four days after kidding.

Kids should be disbudded (have their newly developing horns removed) when they are five to ten days old, and buck kids not intended for breeding should be castrated at the same time; see pages 75–77 for both techniques.

Problems

Kidding problems are rare and chances are you will never have to worry about an abnormal kidding. If your goat is in good health she is unlikely to have any problems. However, if your doe is in labor for thirty to forty-five minutes without producing a kid, assist her as you would a ewe. For a detailed description, see the section on abnormal lambing in Chapter 5.

There are some special conditions associated with kidding that a goatherd should watch for. Ketosis, a pathological accumulation of ketone bodies, can occur a few weeks before kidding; it is likeliest to affect does that are either too fat or malnourished. It is caused by the inability of the liver to transform body fats into sugar, or from an inability of other tissues to meet the carbohydrate demands of the doe and unborn kids; it is commoner in does carrying multiple fetuses. Symptoms include the doe's acting listless and weak, and probably not eating well. Her coordination will become progressively poorer and eventually she will not be able to get up. Ketosis is fatal if allowed to progress. Call the veterinarian if you suspect this problem.

Milk fever can occur six weeks before kidding, but it is generally seen up to ten weeks after the kidding. The condition is characterized by calcium's being used to produce milk faster than the doe's glands replace it. Symptoms are a subnormal temperature and a weakness in the rump and back legs. The doe may drag her back legs as she tries to stand up. She will not be alert and her eyes will appear glazed. Call the veterinarian if you suspect milk fever.

Mild hemorrhaging is common after kidding, so do not be concerned unless it becomes severe. The hemorrhaging will lessen a bit each day until it disappears within a week or two.

A prolapsed uterus occurs when the doe continues to strain after kidding and thus forces the uterus out. The uterus is a heavy-looking red mass with roundish fleshy outgrowths on it. Call the veterinarian immediately. While waiting for the veterinarian's arrival, keep the uterus covered with a warm, moist towel, and make sure that no animal or person steps on it. It is unwise to breed a doe whose uterus has prolapsed, as the chances are great that she will again expel it.

Udder edema can occur either just before or just after kidding. The udder feels like bread dough to the touch and you may see your fingerprints after handling it. The doe may be in pain when you handle the udder. Udder edema is caused by improper blood flow through the blood vessels in the udder. Massage, hot compresses, and bag balm (an antiseptic petrolatum ointment) help restore normal blood flow. If the condition continues for a few days or worsens, call the veterinarian.

FEEDING KIDS

You have the option of letting the kids nurse or of bottle feeding them. Having them nurse eliminates your having to bottle feed the kids, and if your doe has very small teats, nursing will probably enlarge them. Even if you decide to let the kids nurse, milk the doe twice a day to maintain her milk-production level.

The disadvantages of nursing are numerous. It is difficult to determine and regulate the amount of milk each kid receives, and this factor may adversely affect its development. Nursing kids are rough on the doe's udder and may cause damage to udder tissue. Nursing also encourages the development of extra muscle tissue in the udder, making milking more difficult. Kids that nurse seldom develop as strong a relationship with people as those who have been bottle fed, which is an important factor for any goat raised for breeding.

If you decide to feed the kids, start right away; if kids start to nurse it is hard to convince them that milk can come from any other container. You may either pan feed or bottle feed your kids. To pan feed, simply put the milk in a pan and let them drink. You will have to teach them at first by dipping your fingers into the milk and letting the kids suck your fingers. Each time you redip your fingers, lower them closer to the pan, until they are in the pan and the kids are drinking. The problem with pan feeding is that goats, like cows, are ruminants, and have four stomachs. Milk must get to the fourth stomach (the abomasum), which is accomplished by the milk's bypassing the slit in the esophagus and entering the third stomach (the omasum), where it is mixed with digestive fluids and passed to the fourth stomach. If the kid must bend its head down to drink from the pan, milk is apt to pass into the slit in the esophagus and enter the first stomach. The kid may then scour (have diarrhea), which can be a fatal problem.

Kids will assume a natural feeding stance when bottle fed if you hold the bottle at a proper teat angle. The correct stance is head up, neck stretched out and up, front legs spread wide apart, and tail wagging as fast as it can go. The stretching enables the slit in the esophagus to be closed off. To bottle feed, you will need a lamb or kid nipple that fits over a soda bottle or canning jar; the nipples are available at any feed store at a low price.

Do not assume that a kid is born knowing all about nursing. You probably will have to help it get a good start by cupping its lower jaw in one hand and using your fingers, if necessary, to open its mouth. Use the other hand to hold the bottle. It may take several feedings before the kid catches on.

You can feed the kids twice a day, but three or four times a day is best until they are eating hay and/or grain well. The kids' size and vigor determines how much milk to give them; it varies from 2 to 4 ounces per feeding at birth, if you are feeding them three or four times a day. After the doe stops producing colostrum, offer the kids milk three or four times a day at the rate of 4 to 6 ounces per feeding. Gradually increase the amount to $1\frac{1}{2}$ to 2 pints a day. When the kids are consuming a full ration of milk and eating some grain and hay, provide the bottle twice a day.

The milk for the kids must be warmed to 100°F to 103°F (40°C) to approximate the temperature of the doe's milk. Be careful not to scorch the milk, especially if it is colostrum. Offer the kids an equivalent amount of water after their milk. The water should be warmed to the same temperature and fed from a bottle; it discourages scouring and the kids enjoy it.

If you want all the goat's milk for yourself, you may satisfactorily raise the kids with a commercial milk replacer for calves. Feed the kids the colostrum as long as the doe produces it, then feed the milk substitute in the same quantity as milk. Goat's milk is much sought after, and if you have a buyer for it you may find it more economical to use a commercial product to raise the kids.

About a week after birth, the kids will begin to eat hay. Encourage this by making sure they have fresh hay in front of them at all times. Hay is important for the development of their digestive systems. When to start grain feeding is subject to a variety of opinion; you may start feeding calf starter when the kids begin nibbling hay. They are ready for regular goat feed when they are about a month old.

Weaning

The kids may be weaned after three to four months. If they are eating hay and grain well you can wean at ten weeks, particularly if you include a supplement such as calf manna with the grain. The kids should be drinking water from a bucket before you wean them.

Once the kids are weaned, allow them to free feed until they are six to eight months old. The growing goats will consume about 1 pound of grain per day, gradually increasing to a dry doe's ration of $1\frac{1}{2}$ pounds night and morning.

FEEDING GOATS

The nutritional requirements of dairy goats have not been accurately determined, so it is difficult to ensure a balanced diet without commercial feeds. Goats, as browsing animals, can successfully have their diets supplemented by grazing even unimproved pasture and brush areas. In deciding on a feeding plan for your goat, remember that a satisfactory diet is necessary for good milk production.

Commercial feed may be a grain mix with a 12–14 percent protein content if you are feeding clover or alfalfa hay, or a 16–18 percent protein content if you are feeding a grass or grass/legume hay. Goats prefer the coarse variety of grain. Suggested rations for goats are given in the following table:

DAILY RATIONS FOR GOATS

(To Be Divided Between Two Feedings)

	Grain	*Hay*
Bucks	2 lb.	2–3 lb.
Dry Does *(two months before kidding)*	3 lb.	2–3 lb.
Milking Does	$\frac{1}{2}$ lb. of grain for each 1 lb. of milk produced	2–3 lb.
Growing Goats *(from six months until two months before kidding)*	1 lb.	2–3 lb.
Kids	Free feed first six months	

The hay quality determines the protein level necessary for the grain ration. Remember to feed hay from a hayrack to minimize waste. Alfalfa and clover hay are the best for increasing milk production.

Goats need a constant supply of fresh water. Keep the water container clean or the goats will not drink from it. Water requirements vary considerably, but plan to offer at least $1\frac{1}{2}$ to 2 gallons a day per milking doe and increase or decrease according to need.

In addition, you may provide a salt block for your goats, although minimum requirements for salt should be met by the commercial feed. Milking does may be offered some dried citrus rinds. These can be bought at feed stores and is believed by some to improve milk flavor subtly; goats really enjoy it.

Pasture

If you have sufficient space to let your goats go out to pasture, check the area for poisonous plants. The U.S. Department of Agriculture has a list of poisonous plants indigenous to your area. Some common plants that are poisonous to goats are bracken fern, European hemlock, laurel, loco weed, and wild cherry. Do not worry about poison ivy; goats love to eat it. However, make sure you wash the udder well before you milk, or your hands may be itching shortly thereafter.

Your pasture area should provide some type of shade for your goats. If natural shade is not available, build a simple shed with a few posts and boards. Make your structure sturdy, however, as your goats are apt to climb up on the roof.

Pasture areas for goats should be rotated at frequent intervals to avoid a parasite buildup. A pasture area may be divided into two sections with the goats rotating every two to six weeks, depending on the size of the pasture and the number of animals grazing on it.

COSTS

Determining the cost of maintaining a milch doe for one year is highly satisfying when you look at the value of the products produced. Our example assumes one goat producing a low average amount of milk for a 305-day lactation period, and a twin kidding. The value we have placed on milk is based on a 1981 average rate for an equivalent amount of cow's milk. The sale price for kids is an average meat price and is therefore considerably lower than the price of kids sold for breeding, particularly if you have registered stock. The purchase price for the doe is typical for a quality grade goat at a prime age; if you buy a registered goat at a prime age, the cost is higher. The original purchase price is divided over the years of expected productivity, and therefore our calculations include one-sixth of the doe's purchase price in the annual cost of maintaining the animal.

COST OF MAINTAINING A MILCH DOE

Purchase Price	
$75 ÷ 6	$ 12.50
Feed	
830 lbs. of grain @ 10¢ per lb.	83.00
1,120 lbs. of hay @ $2 per 35 lb.	64.00
Breeding fee	15.00
Total Maintenance Cost:	$174.50

COSTS OF RAISING KIDS

(Assumes kids sold at 3 months)	
Feed	
⅓ lb. of grain per kid per day @ 10¢ per lb.	$ 4.20
⅓ lb. of hay per kid per day @ $2 per bale	6.00
Total Raising Cost:	$10.20

Product Value	
Milk	
1,250 lb. milk (305 lactation days)	
− 219 lb. for raising kids	
1,031 lb. = 128 gallons of milk @ $1.65 per gallon	$211.20
Sale	
$35 per kid	70.00
Total Product Value:	$281.20

Profit	
Total Produce Value	$281.20
− Total Cash Costs	184.70
Total Product Benefit	$ 96.50

These costs and product values are approximations. The cost of your doe can be significantly lowered if there is pasture or garden surplus to supplement the diet. The product value increases if some of the milk is used to produce more costly items per pound, such as ice cream or cheese. And though we used the price for an equivalent amount of cow's milk, goat's milk is considerably more expensive: a quart of goat's milk costs from 75¢ to $1.25, depending on supply and demand.

CHEVON

Goat meat, called chevon, tastes like lamb, but it is milder and less greasy. Goat kids weighing twenty to thirty pounds bring a premium price during the Easter-Passover season, when goat meat is often used for holiday meals. You may dress out (clean for cooking or sale) a kid at birth just as you would a rabbit; older goats may be cut into roasts and chops. A buck headed for the table should be castrated about two months before dispatching if it is over three months of age. Old goats may be ground up for hamburger.

It is difficult to think of eating chevon when you have been bottle feeding the kids, but it is a far better fate for bucks than ending up as an abused pet. We never have been able to eat one of our own kids, but we have happily bought and eaten other people's.

MAINTENANCE

Goat's hoofs need to be trimmed several times a year. To neglect this chore is to invite lameness and ultimately contracted tendons, deformity, and foot rot. If your goat's hoofs are long overdue for a trim, it may take several trimming sessions to get them back into shape.

Select a sharp jackknife, a linoleum knife, small pruning shears (our choice), or hoof nippers. Stanchion your goat in the milking stand with some grain, or have another person hold the goat. Trim the hind hoofs first. Stand with your back to the goat's rear or put one of your feet on the milking stand and bend the goat's leg at the knee over your knee. Clean out the accumulated dirt and manure on the hoof. Then trim off any portion of the hoof that is folded over. Next, trim the remainder of the hoof until the white part takes on a pinkish color. The wall of the toe should be level with the cushion of the hoof.

Do the front hoofs in the front manner. The most comfortable position is to bend the goat's front leg at the knee and bring it under the goat's body. When you are done trimming, your goat should stand squarely on its feet.

Brush your goat with a coarse brush before milking. Trim any especially long hairs, particularly around the udder and flank. Brushing removes loose hair and dirt, and helps keep away lice.

You may bathe your goat with warm, soapy water. Rinse with warm water after the bath and dry the goat's coat well. Do not expect your goat to like the bath.

Castration

Unless you definitely plan to use a buck kid for breeding, you should castrate him at an early age. You can castrate as soon as the testicles descend into the scrotum, which generally happens in the first three weeks. The earlier the job is done, the less shock there is to the kid.

The castration method most commonly used by homesteaders requires a razor blade or a sharp knife. Rinse your equipment in a mild disinfectant before use. Have somebody else hold the kid on his or her lap while you perform the operation.

Wash the scrotum with warm, soapy water and dry it. Next, hold the scrotum between your thumb and forefinger, causing the testicles to move toward the kid's body. Make an incision about an inch long at the end of the scrotum. The testicles, covered by a white membrane, will be visible. Pull one testicle well out of the scrotum. If the buck is very young, keep pulling the testicle out until the sper-

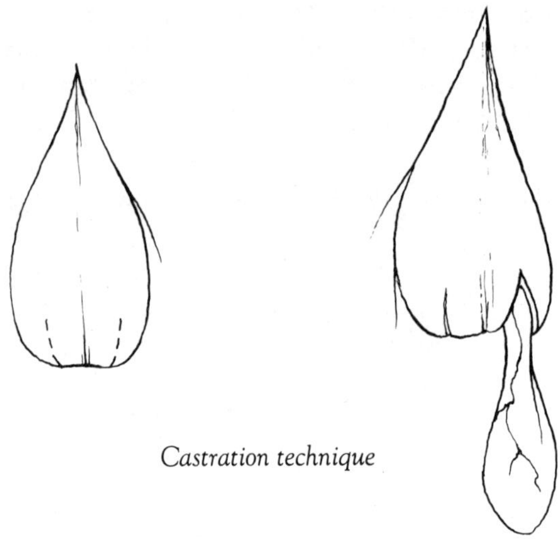

Castration technique

matic cord snaps. If the kid is more than a few days old, scrape the cord with your cutting instrument until it breaks; if you simply cut the cord there is apt to be much bleeding. When you have removed one testicle, sever the other in the same manner. When you are finished, dose the wound with an antiseptic. Keep checking the kid until the wound heals to make sure infection does not set in.

A second castration method uses a Burdizzo emasculator. This is preferred because it leaves no wound to become infected; its disadvantage is the expense of the tool. It is better to use this method on older goats, as the older the animal is, the bloodier a knife castration. After locating one testicle in the scrotum, place the emasculator across the cord, being careful not to include the scrotum's central division. Close the emasculator's handles and keep them closed for ten to fifteen seconds. Repeat the process on the other testicle. The crushed cords will cause the testicles to atrophy.

A third method involves an elastrator, a device with a rubber ring or rubber bands. Place the rubber ring or band around the scrotum, making sure the testicles are below the ring. In time, the testicles will atrophy. As this method sounds so simple, it should be stressed that it is painful and may easily produce an infection; there is also the risk that the elastrator could slip. The other two methods are quicker, surer, and much safer.

Dehorning and Disbudding

If you own a mature goat with horns you may have it dehorned; this requires the skill of a veterinarian. Generally speaking, dehorning is expensive and arduous, and veterinarians are none too eager to do it.

Disbudding, on the other hand, entails destroying the horn buds before the horns start growing, and can be done quite easily by the homesteader. Plan to do it when the kids are three to sixteen days old. There are two methods of disbudding: using a disbudding iron and using dehorning paste. Of course, a veterinarian may do the job for you.

The disbudding iron is relatively inexpensive and easy to use. Heat the iron until it is hot enough to burn a perfect circle on a piece of wood without much pressure. While the iron is heating, trim the hair around the goat's horn buttons (the tips of the emerging horns). Have a second person put the kid across his or her lap and firmly hold its muzzle, then press the iron over the button. Rotate the iron slightly to make sure you are applying it evenly all around the horn button. The kid will struggle and may scream, but in ten to fifteen seconds it will be all over. Reheat the iron and do the other horn button the same way. You will feel better if you have a bottle of warm milk to give the kid when you are done.

If you did not have the iron hot enough or did not hold it on long enough, you may see scars—thin, misshapen horns—develop after a time. If this happens, heat the iron and redo the job.

Dehorning paste burns the horn buds. Trim the hair around the horn button, cover the horn buttons with adhesive tape and put petroleum jelly all around the skull. Remove the tape and apply the paste to the horn buttons. After applying the paste, keep the kid separated from other kids for a day or two, so they do not lick the paste, which is highly caustic and can cause blindness or severe burns. Do not let a kid treated with the paste nurse for a day or two, as the paste will burn the doe's udder.

MEDICAL CONCERNS

It is far easier to keep a goat healthy than to cure a sick one. The following simple rules will contribute to maintaining healthy goats:

- Do not subject your goat to stress: a rapid change in diet, lack of water frequently or for long periods of time, interference from other animals, and extremes in weather. In cold weather, be sure that the goat quarters are dry and draft-free and that the animals have access to fresh air. In hot weather, see that they have shade in the exercise yard and plenty of fresh water. Also make sure that the sleeping quarters have as much ventilation as possible.
- Keep the goat quarters and water and feeding pails clean. Remove any nails, fencing, or other objects that might injure the animals.
- Store feed appropriately.
- Provide an adequate diet.
- Check the overall appearance of your animals daily.

To avoid chaos if a medical problem develops, set aside a place for the goats' medicine chest. Make sure that it is out of reach of the children. Keep the following useful items in the chest:

- A veterinary-strength antiseptic, such as iodine or boric acid
- Scissors (for trimming hair)
- Worming pills
- Germicidal soap and/or sterile gloves
- Mineral or vegetable oil
- Clean cloths or towels
- Hoof-trimming equipment
- Udder balm
- Veterinary-strength liniment
- Thermometer
- Astringent powder

Although goats are among the healthiest of domestic animals, there are a few diseases with which you should be familiar. Abscesses are common among goats, and in most cases will not cause any problems. If an abscess is draining, keep the pus cleaned off, treat with iodine and isolate the goat to prevent the bacteria that cause the abscess from spreading. Dispose of the material you use to clean off the pus and wash your hands thoroughly each time you touch the area. If the abscess is on the udder, do not drink the milk until the abscess has cleared up.

Brucellosis (Bang's disease or Malta fever) is caused by bacteria that are spread primarily by ingestion. Milk from a dairy animal with brucellosis can cause undulant fever in humans. Although the disease is very rare among goats in this country, you should have each goat you add to your herd tested for brucellosis by a vet.

Bloat can be caused by overeating grain or new pasture. Preventive steps can help avoid this. Do not turn your goats out on a new pasture until their digestive systems have adjusted to that type of food. This is particularly true with fields of rapidly growing leguminous plants such as alfalfa and clover. You can bring some of the pasture feed to them for a few days and feed them hay each day before you turn them out to the pasture. Turn them out for only several hours at first, making sure that the dew is off the grass; wet pastures are more likely to cause bloat than dry fields. The first visible symptom of bloat is a distension on the left side; breathing may be difficult and there may be profuse salivation. A dose of vegetable or mineral oil (about a pint) may be administered while you are waiting for the veterinarian to arrive.

Foot rot is an infectious disease of the feet caused by a fungus, and is most common when goats spend long periods on wet ground. The first symptom is lameness. To treat foot rot, trim the hoofs properly, soak the affected feet in a solution of Epsom salts, and rinse with an astringent. Do this a couple of times a day until the condition clears up. Be sure to keep the affected animal on dry ground and bedding while being treated.

Indigestion is indicated by the goat's stamping its hind feet, making grunting noises and poking at its side. There will probably be some signs of diarrhea. If indigestion can be anticipated—as when Lola found the feed barrel—give the goat a dose of four to eight ounces of vegetable or mineral oil. If the goat is already suffering from severe indigestion, call the vet.

Mastitis is an inflammation of the mammary glands due to the presence of bacteria. There are three types: acute, chronic, and gangrene mastitis. Acute mastitis symptoms are a fever and a hot, hard, swollen udder. The goat will often act as if she is in pain and her milk will contain chunks or ropy mucus. Chronic mastitis is detected by hard lumps in the udder and abnormal milk for a few days. The goat will probably not act sick, although occasionally she may show signs of acute mastitis. Gangrene mastitis can come on very quickly; it is characterized by a cold, bluish discoloration of the teats and/or udder. Do not drink the milk from a goat you suspect of having mastitis. To treat the problem, call the vet.

The main symptom of pneumonia is a high temperature, 104°F to 106°F (42°C). The animal may act sick and show signs of gasping for air, or may not. Isolate the goat (some types of pneumonia are contagious) and call your vet.

Poisonous plants ingested will cause vomiting, frothing at the mouth, a staggering gait, and possible convulsions. If the symptoms are not too severe, a teaspoon of bicarbonate of soda in the goat's mouth helps by increasing vomiting. If the symptoms are severe, call the veterinarian.

Scours (diarrhea) can be fatal to kids. To prevent scours, be sure to warm the milk and not to overfeed. If you notice diarrhea, give only half the amount of milk you have been giving when you feed the kid. The problem should clear up within two more feeding periods and you can increase the amount of milk at a slow rate. Watch for any recurrence of diarrhea. If cutting back the amount of milk does not clear up the problem by the next two feedings, you may give a dose of a commercial antidiarrheal agent with the reduced portion of milk. If the problem continues, call a vet.

Tuberculosis symptoms are general weakness, emaciation, a low-grade fever and sometimes an accompanying hacking cough. Goats in this country are free of tuberculosis; in fact, there apparently has never been a reported case of it here. However, milk from a tubercular goat could be a transmitting agent for the bacilli to humans, so you may want to test any new goat added to your herd. The test can be administered by a veterinarian at the same time you have the animal tested for brucellosis. If you sell milk, most state laws require you to have the animal retested for tuberculosis and brucellosis every two to three years.

Worms are a perennial problem and there is considerable difference of opinion about worming goats. Many goat owners worm their animals twice a year with a broad-spectrum wormer. Periodic fecal-sample testing by your vet is a good idea, as then you may be sure that you are treating for the type of worms present. Also request that the vet check for coccidiosis, a disease resulting from an infection of the digestive tract by parasitic protozoa. It can prove fatal.

Wounds can be prevented by removing sharp objects from the shelter and exercise yard. Still, it is a good idea to give your goat a tetanus booster on a yearly basis. Treat any wound with an antiseptic. If the wound is deep, apply pressure with a damp cloth for five or ten minutes. If the bleeding does not stop, call your veterinarian. This procedure is sufficient for surface wounds, even those to the udder, which tends to bleed profusely. Keep checking the wound until it heals to make sure infection does not set in.

It is no wonder that the goat has been dubbed the "Depression animal": it is an efficient and effective supplier of protein. Goats are also easy to care for and have a long history of symbiotic relations with humans, being one of the first domesticated animals. More people around the world drink goat's milk than cow's milk. Goats are easy to transport in the back seat of the family car, a very important consideration for a homesteader, and they require relatively little space. Last, but of major value, they are exceedingly loving. We strongly recommend a doe as the homesteader's first animal.

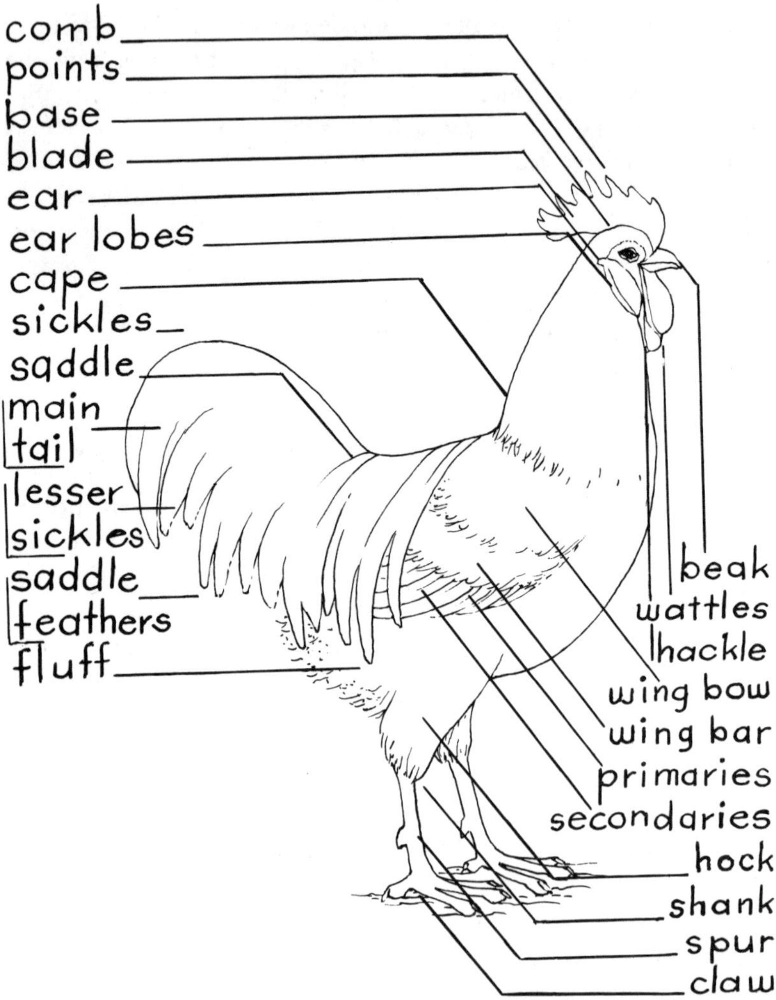
comb
points
base
blade
ear
ear lobes
cape
sickles
saddle
main tail
lesser sickles
saddle feathers
fluff
beak
wattles
hackle
wing bow
wing bar
primaries
secondaries
hock
shank
spur
claw

4 CHICKENS

CHICKENS are undoubtedly the most common homestead animal, and with good reason. Aside from their well-known efficiency as egg and meat producers, they are easy to care for, inexpensive to buy and maintain, and do not require much space. The feed:egg:meat conversion ratio is 4.5:1 per dozen eggs and 3:1 per pound of meat.

Caring for chickens and the eggs is an activity that can be easily handled by young children, once they understand a few rules about not chasing or yelling at the birds. Of all the chores in which we involve visitors, egg collecting seems to give them the greatest pride. Our own children have long since gotten over being thrilled at the chore, but still delight in collecting the money they earn when they sell their eggs.

One fall we wanted to buy a dozen replacement hens and we located a retiring poultryman who was selling his birds for a good price. Both price and birds were so good that we figured we had at last found an excellent use for the unused space on the second floor of our barn. We bought 175 hens and we were in the egg business.

We stayed in business until the following July and then sold out in a week to avoid housing them through a hot and humid summer. I asked everybody in the family not to sell my favorite chicken, but unfortunately she was also the most friendly, so she was sold with the first dozen hens. Talk about being "mad as a wet hen"!

The only reason we did not go back into business on that scale was because we enjoyed our egg customers too much. We had forty-four customers per week and the average time spent with each customer was twenty minutes. We loved every minute of the people sharing their lives both past and present with us, but unfortunately homesteads do not run on good conversation alone.

The adult female chicken is called a hen and the adult male is a cock or rooster. A female under a year is a pullet, a young male is a cockerel, and a baby chicken of either sex is called a chick. A capon is a cockerel whose sex glands have been removed. Caponizing allows the bird a long growth period before it is used as meat. The procedure also improves the quality of the meat because the amount of fat is increased.

When you order chicks, you must choose sexed or straight run. Sexed chicks are separated by gender; straight run means an approximate fifty-fifty mix of each gender, the way they come from the incubator.

You also have a choice of age. You may purchase day-old chicks, started pullets (six to eight weeks old), or ready-to-lay pullets (about five months old). The price increases with the age. Second-year layers can usually be picked up for a good price in the fall; these birds have had their first annual molt (change of feathers) and therefore tend to lay fewer (by about 10 percent) but larger eggs. We say tend to, because we have had Reds and Black sex-links that have laid jumbo eggs the first year and continued to lay five to six eggs per week during their second year.

You may see reference to the national breeding grades when buying chicks. These grades refer to the egg-laying ability of the chicks' ancestors. The grades are as follows:

- U.S. Record of Performance: mother and paternal grandmother were both at least 200-egg (per year) producers
- U.S. Certified or U.S. Verified: offspring of good hens and a rooster whose mother was a 200-egg producer
- U.S. Approved: offspring of excellent examples of the breed involved

Chickens are given a blood test for pullorum, a bacterial infection transmitted from hen to chick; it can kill as many as three-fourths of the chicks in a hatching. In tested flocks, the positive stock are removed. Flocks tested are graded as follows:

- U.S. Pullorum Clean: no sign of the disease on two testings at least six months apart
- U.S. Pullorum Passed: no sign of the disease on one testing done during the year prior to sale of stock
- U.S. Pullorum Controlled: fewer than 2 percent of the stock had the disease
- U.S. Pullorum Tested: fewer than 7 percent of the stock had the disease

BREEDS

There are more than two hundred breeds of chickens and all come in two sizes: standard and bantam. Bantams weigh approximately one third or less than the standard-sized birds. Standard size refers to the full size of that particular breed; not all full-sized birds are one size. For example, the standard-size Rhode Island Red hen's average weight is seven pounds, whereas the standard-sized Leghorn hen is four pounds. Chicken breeds vary in shape, color, temperament, egg-shell color, feed efficiency, and a number of other areas.

The American Poultry Association has developed a "standard of perfection" as a guide and system for listing and describing the various breeds. Purebred chickens are first classified according to class, indicating the original geographical area of breeding. They are further classified according to breed, which gives an indication of general shape and other physical features. The final classification is varieties, indicating the color pattern or comb type found within a given breed. Using the standard of perfection, the Plymouth Rock is classified as follows:

Class: American
Breed: Plymouth Rock
Varieties: Barred, Blue, Columbian, Partridge, Silver Penciled, White-Buff

SELECTING CHICKENS

Most homesteaders are interested in a good dual-purpose breed or strain of chickens (dual-purpose chickens are both good egg layers and meat birds). The most popular are the Rhode Island Reds, Plymouth Rocks, New Hampshires, and various strains of these breeds. The hens are medium-sized, weighing $6\frac{1}{2}$ to $7\frac{1}{2}$ pounds.

Before you buy chickens, decide what color eggs you want. If you live in New England and plan to sell eggs, you want brown eggs; in the rest of the country, white eggshells are more popular. The shell color does not affect taste or food value of the egg, but local preference affects the market value.

Examples of brown-eggshell layers are Rhode Island Reds, New Hampshire, and the Rocks. Leghorns, Anconas, and Minorcas are white-eggshell layers. The Leghorn is the most popular and efficient egg layer; it is lightweight, with the hens weighing 4 to $4\frac{1}{2}$ pounds at maturity. For variety, the Araucania or Easter Egg chicken lays eggs with blue or blue-green shells.

In addition to purebreds, there are a number of strains resulting from crossbreeding. The strains frequently produce offsprings with sex-linked colors. For example, Barred Rock hens mated with Rhode

Island Red roosters will produce barred males and black females with gold lacing on the neck, whereas a Barred Rock male and a Rhode Island Red hen will produce barred chicks of both genders. The strains generally are developed for high egg-meat production, but do not produce offspring of predictable characteristics.

The best meat chickens are the hybrid Cornish-White Rock crosses used by the commercial poultry industry. Breeds commonly used for meat production are White Rocks, White Wyandottes, and New Hampshires. Broilers are young chickens, eight to ten weeks old, that are used for broiling or frying; roasters are three to five months old and weigh five pounds or more; capons are six to seven months old and weigh seven pounds or more; and stewers or fowl are old roosters or hens that have past their peak egg-production period. The age and weight of fowl depend upon the breed.

Hens will lay eggs with or without a rooster, although a rooster is needed if you want fertilized eggs (and more chickens). A light-breed rooster can service about twenty hens and a heavier-breed rooster can handle ten to twelve hens. Having a rooster and an appropriate number of hens does not, however, ensure that every egg will be fertilized. The length of fertility for roosters varies greatly, but generally they are fertile as long as they are sexually active.

Roosters are beautiful to look at and fun to watch, but roosters can also be mean and attack people; they can be dangerous, especially to young children. An attacking rooster uses his beak and wings, but his primary weapons are the long spurs on the lower back of his legs.

Chicken breeds can be mixed successfully if the breeds are of similar activity patterns and temperament. For example, the active, nervous Leghorns are more suited to mixing with Minorcas than with Reds or Cochins. Crested birds such as the Polish tend to have the least dominant personalities. Your best bet is to have egg producers of one breed and some pet range chickens for fun and variety. Chickens with free range and a lot of space can be mixed much more readily than confined layers.

Mixing ages can be very unsatisfactory. Do not put birds under five months with mature birds. The older birds will dominate the younger ones, may injure them and may transmit diseases for which the young birds have not yet developed resistance.

Chickens fight when new members are introduced to the flock because they have established a social order (pecking order) among themselves which new members disturb. Before introducing new members to your flock, fence them next to each other for a week or so; the fence keeps them separated, but allows them to get acquainted. When you do finally put them together, stay and watch the flock and remove any chicken that is not accepted.

Once we picked up a nice young six-month-old Rhode Island Red rooster for our Red hens. We thought he was gorgeous, but the hens did not like him at all. To save his life we had to fence him separately for a couple of days. After that, they still gave him a hard time occasionally, but they did allow him to be with them. He became the boss in about six months.

SHELTER AND EQUIPMENT

Chickens' housing requirements are largely determined by your needs. If you plan to keep chickens all year, have your coop electrified or your chickens will stop laying when daylight lasts less than fourteen hours. If you plan to keep chickens for the six warm months and use natural light, a portable range house in a fenced area is sufficient.

We like to keep our manure detail down to a minimum, so we use a portable house all year round. We drag it into the barn in the fall and move it back outside in the spring. Our chicken house used to be a kennel for two pug dogs. One end is an enclosed house with a small door, which is where we placed the nesting boxes. The run (screened exercise area) is four-by-eight feet, with the sides and floor made out of heavy wire mesh. The roof is made out of exterior-grade plywood with the edges slightly extended. It stands on corner two-by-four-inch legs that are extended to raise the house a foot off the ground. We nailed feed sacks along the north side to keep out wind and rain.

CHICKENS

The chicken coop should provide clean and dry quarters all year round. The ideal temperature range for the coop is 45° F to 80° F (8° C to 27° C) for maximum egg production. An area of your barn, garage, or a small existing outbuilding can be used.

Chickens give off a lot of moisture, which can be a problem. If your coop is insulated you will need a vapor barrier between the insulation and the interior wall. The coop will also need a good ventilation system to control the moisture without drafts; a couple of windows opened at the top can be very effective.

The biggest disadvantage of chickens is that they are extremely dirty compared to other homestead livestock. One chicken produces more than two hundred pounds of manure per year. Choose a solid-floor coop designed for easy cleaning. Allow enough head space for you to be able to stand up while shoveling. If the hens are housed in a part of the barn, it is easier to clean the area if it is near a door or window.

If you are using a portable chicken house outside, make the floor out of a heavy wire mesh so that the droppings will fall through to the ground. Move the house when the manure builds up.

The size of the coop depends on how many chickens you have in your flock. Laying hens need about 2 to 3 square feet of space per bird. If you live in a hot climate, $3\frac{1}{2}$ square feet per bird is even better. In a cold climate, hold the space per bird to the minimum so that the chickens' body heat helps maintain a satisfactory temperature during winter months.

Laying hens need to have roosts. The roost may be made from a two-by-four-inch board placed narrow side up and the top two edges beveled, or from a well-cured branch. Lightweight breeds need eight inches of roost space per bird; large breeds require about ten inches. Place roosts twelve to fifteen inches apart at an angle to the ground, the highest toward the back.

There will be a large accumulation of droppings under the roosts, so design it with ease of cleaning in mind. A little chicken wire stapled to keep the chickens from going under the roosts will help keep their feet clean.

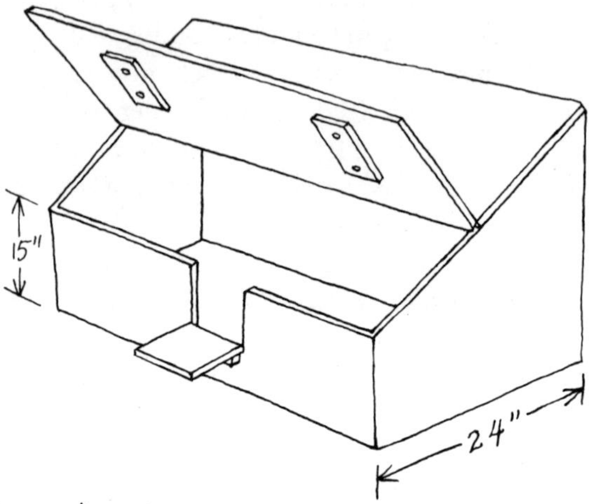

Community nest

You need to provide nests for your laying hens. They may be built out of metal or wood. A community nest is one big enough to accommodate several hens at one time. This type of nest should provide one square foot for every five hens. The interior of the nesting area should be darkened, so enclose the nesting area, with the exception of an eight- to ten-inch entrance. Install a hinged roof or back so you can get to the eggs easily.

Individual nests are usually ten to twelve inches wide, twelve to fourteen inches high, and about twelve inches deep. Plan to provide one nest for every four hens. Put the nest boxes about three feet above the floor and build a perch below the entrance to help keep the nest clean. Provide plenty of clean straw or shavings to help keep the eggs clean.

It is not necessary to provide a chicken yard, but it is fun to have one. Allot however much space you care to. A five- or six-foot-high fence of heavy enough material to keep out cats and dogs is required. The bottom of the fence should be stapled to a frame or staked to the ground to keep the chickens in and predators out. The yard does not have to have a wire top, but if it does not, be sure to clip one wing of each chicken so none flies out.

Ideally, the yard should be on the south side of the coop so that it can be used all year. The coop door should be locked at night to keep out nocturnal predators. Provide a ramp from the door to the ground.

The floor of the coop needs some type of litter. The best litter material is straw, but if availability and cost make it prohibitive, you can use shavings, sand, ground corncobs, or peat moss. If you plan to use manure as a fertilizer in the garden, choose peat moss or straw.

You may spread a thin layer of litter and shovel it out every few days, but this is both expensive and time consuming. The deep-litter method, also called compost litter, is the most practical for the homesteader. Start with a clean, dry floor and fully cover it with four to six inches of litter. Stir the litter up at least once a week and add a thin layer of new litter until the litter is eight to twelve inches deep. The litter is stirred to keep it from caking and to mix the fresh droppings with the material below; a natural bacterial action sanitizes the mixture.

The deep-litter system keeps the litter dry on top. Add a new light layer of litter when necessary. Any wet spots, such as will be found around the waterers, should be cleaned out as needed and fresh litter put down. During cold, damp weather, scatter some hydrated lime over the surface before stirring to help keep the litter drier.

The litter is cleaned out in the fall and/or spring of the year and a new deep litter started. Our experience has taught us to clean out the litter on a cool day and to get all the help you can round up. It is a dirty, dusty job, but it can be over in a couple of hours if everybody pitches in. Bandannas worn over the face is a good idea.

Chicken feed may be thrown on the ground, but the amount of feed that will be wasted makes this a poor method. There are many types of chicken feeders, all of which help cut down on wasted feed.

The two most common types are the metal hanging feeder and a V-shaped trough. One hanging feeder fifteen inches in diameter provides enough feeder space for fifteen hens. A three-foot V-shaped trough provides sufficient space for fifteen layers, as they need four inches of feeder space per bird.

You can make a fine V-shaped trough out of scrap lumber. Fashion a reel (or string wire attached to a door spring) across the top to keep the chickens from messing in their feed. A wire and door spring allows you to fill the trough without wasting feed.

Design your feeder so that its top is raised to at least the level of the birds' backs as they stand on the floor. Putting a lip on the side of the feeder also helps prevent the chickens from wasting feed. Fill the trough feeder about halfway as additional protection against wasted feed.

Unless you are feeding a complete laying ration, chickens also need oyster shells (for calcium) and grit (as a grinding agent), and these require a separate feeder. You can nail a one-by-four-inch board between the studs of the coop to dispense these items. You may use the chickens' own eggshells as a calcium source, but crush the shells to small pieces so as not to encourage the hens to eat their own eggs.

Chickens require ample fresh, clean water. A one-gallon waterer will provide enough water per day for twenty hens. Waterers come in varying sizes and styles; choose a sturdy waterer, as the less water spilled, the less litter you will have to change. If you live in a region that gets cold enough for frozen water to be a problem in the winter, consider a waterer with an electric water heater.

Hens need a minimum of fourteen hours of light per day to produce eggs efficiently. The light stimulates the pituitary gland and this produces greater laying. It also gives the hens more time to eat and therefore to produce more eggs. To maintain egg production through the shorter days, you will need artificial light. Use a fifteen-watt bulb with a reflector hung three feet off the floor in the center of the room. Turn the bulb on in the late afternoon and leave it on all night. You can buy or make an automatic switch and set it so the lights go on at 4 A.M. Use a twenty-five-watt bulb with this system.

FEEDING

The higher the feed consumption of hens, the higher the egg production, so you want to do all you can to encourage them to eat. Protect the feeders so the chickens cannot mess in their food and provide fresh feed every day. It also helps to run your hands through the feed in a trough when you gather the eggs; the chickens think you have added more food, and so they eat more.

It does not make any difference what time you feed and water your hens, but be consistent once you pick the time so that they are never out of feed for more than a half hour. Morning and late afternoon are good feeding times, as it ensures that they will have feed in the early morning, before you get out to the coop.

There are several systems for feeding chickens, all of which supply a sufficient and adequate diet. You may, of course, supplement any of the diets with feed from the garden.

Mash is a ground and well-mixed variety of grains and supplements. There are commercial mashes for chickens of every age and purpose, the primary difference being their protein content. For growing or laying hens, choose a complete mash that requires no grain, grit, or oyster-shell supplement, or a mash that is a supplement to scratch grain. You have a choice of regular ground mash or pelleted mash. The pellets are more expensive, but there is less waste with it.

Scratch is grain fed to the chickens to balance their rations; the best scratch grains are corn, oats, and wheat. The grains must be cracked or ground and are added in amounts equal to the mash. A typical mixture is 50 percent mash, 25 percent cracked corn, and 25 percent crimped oats. The term *scratch* is derived from the practice of throwing the grain on the litter and letting the chickens scratch for it. This system keeps the chickens busy but tends to be wasteful; we use a feeder. Do not feed scratch to chickens until they are about eight to ten weeks old.

Oyster shell and grit are both needed if you feed anything but a complete ration. If your chickens are on pasture, however, they may not need grit.

Pasture can be used to supplement your chickens' diet. The best chicken pasture is clover or alfalfa. If you use pasture, rotate it each year to avoid heavy parasite infestation.

The quantity of feed consumed by chicks tends to increase until they reach their maximum rate of growth, when it decreases slightly. It takes about 22 pounds of feed to raise a chick to twenty-two weeks of age. A laying hen will consume $\frac{1}{5}$ to $\frac{1}{3}$ of a pound of feed per day, depending on her size and laying ability. The type of mash is determined by the age of the chickens and the protein level of the mash. The protein requirements for the various ages are as follows:

Baby chicks (under eight weeks)	18–22% protein
Chicks (8–14 weeks)	15–18% protein
Chicks (14–20 weeks)	16–18% protein
Laying hens (over 20 weeks)	16–18% protein

There are also special mash rations for broiler and breeding flocks.

EGGS

Most hens are ready to start producing eggs when they are from twenty to twenty-five weeks old. Hens are born with thousands of tiny germ cells, each a potential egg. After about twenty weeks, the eggs begin to form into a line in the ovary. A few eggs at the beginning of the line are surrounded by a small cell and begin to grow, forming a follicle. Yolk material is produced in the liver and carried to the ovary through the bloodstream, where it is taken by the follicle.

About every twenty-five hours the pituitary gland signals the release of an egg from its surrounding follicle into the oviduct. It takes about six days for an egg to go from this point to the nest. Along the way, egg white (albumin) is picked up, a soft inner membrane is added, and finally the eggshell forms. During these six days the egg can be fertilized.

You can tell if an egg has been fertilized by examining the blastoderm (the small white lump on the yolk). If the blastoderm is organized, rounded, and has a translucent spot in the center, the egg is fertilized. An unfertilized egg will be opaque, disorganized, and irregularly shaped. Any other way of telling if an egg is fertilized is inaccurate.

The egg, all formed, waits in the uterus until it is expelled through the hen's vent. The passage through the vent takes only a few minutes. The number of eggs a chicken lays per year depends primarily on her breeding, care, and diet. A good laying hen from an egg-producing breed should produce from 220 to 280 eggs a year, an average of 4 to 5 eggs per week.

Collecting eggs can be one of your more pleasurable chores, providing the eggs are clean. Provide plenty of clean nesting material and sufficient nesting space for your hens. The more often you collect the eggs, the less chance for them to get dirty or broken. Ideally, collect eggs three times a day.

When collecting, use a basket or pail with shavings at the bottom. An "official" egg basket is made of plastic or rubber-coated wire. Every now and then, one of us tries to collect a few eggs using a jacket pocket as a basket and every once in a while the system is successful—but not often.

Poultry-supply catalogs show some fancy egg-washing machines and detergents, but you should not need them. If an egg is a little soiled, clean it by lightly buffing it with steel wool.

After you have collected the eggs, cool them in the refrigerator. The ideal temperature is approximately 55° F (13° C) with 70 percent relative humidity. Eggs handled properly will maintain a quality and taste comparable to store-bought eggs for five to six weeks. High temperatures are extremely damaging to eggs, so collect the eggs more often during the summer months. Eggs cool down much faster in open-wire baskets.

If you are going to sell eggs, you need to know about grading and sizing. Grading refers to the quality of the egg and is done by candling. You can buy a candler or make one by cutting a hole a little smaller than an egg in the end of an oatmeal box and inserting a twenty-five-watt bulb. Turn an egg in front of the hole and look for blood spots or other defects. Small blood spots do not make an egg uneatable, but they usually do not please your customers. Candling also shows the size of the air cell, which indicates freshness. Fresh eggs have a small air sac, which is why you have trouble peeling fresh hard-boiled eggs. Since egg salad is high on our list of preferred foods, we age eggs especially for hard boiling. We hold the eggs for a couple of weeks so that the air sac expands sufficiently to allow easy peeling. (This gives you a clue to the accuracy of the slogan "farm-fresh" found on egg cartons at the local grocery store.)

Eggs are sized with an egg scale and measured in ounces per dozen; eggs are classified as jumbo (thirty ounces and up), extra large (twenty-seven to thirty ounces), large (twenty-four to twenty-seven ounces), medium (twenty-one to twenty-four ounces), and small (under twenty-one ounces).

You may eliminate the grading and sizing process by selling your eggs as "unsized and ungraded." The eggs must be clean and uncracked. It is good business to offer money back or a replacement egg if any egg is not to the customer's satisfaction. Although it is illegal to use second-hand egg cartons unless you cover up the grade and name of the previous user, you may sell the eggs without cartons, or buy unmarked egg cartons from a farm-supply store.

No two eggs, even from the same hen, are exactly alike. Eggshells of brown-egg layers tend to be darker at the beginning of their egg-laying season, and in general the first eggs tend to be smaller.

Occasionally you might have a hen that lays an egg with rough calcium deposits on it. This tends to occur when a hen has been in production for ten to twelve months, and it does not affect the quality of the egg.

Hens sometimes lay a soft-shelled egg. This can be caused by a lack of calcium or vitamin D, or by some diseases. If it happens only once, it may have been caused by the hen's being badly frightened the night before laying the egg. You may also have an occasional egg with a thin egg white. This can be caused by extremely hot weather and improper egg storage. Blood spots may appear in an egg; this is caused by minor hemorrhages that occur along the hen's oviduct.

BROILERS

If you are buying day-old chicks for egg layers, you may as well purchase straight-run chicks and raise the cockerels for meat. The chickens will be ready to sell or for the freezer by eight to ten weeks, depending on weight.

Equipment and space needs are the same for both. Provide $\frac{1}{2}$ square foot per bird until they are two weeks old, and 1 square foot per bird until they are about ten weeks old. If you keep them longer than that, they should have about $2\frac{1}{2}$ square feet apiece.

Broilers need a high protein feed (20 to 24 percent) for the first six weeks. From six weeks on, feed a finishing mash that has a reduced protein level and an increased caloric level. Many broiler raisers provide lights twenty-four hours a day to encourage maximum eating; a sixty-watt bulb is sufficient for 200 square feet.

BUTCHERING

The easiest way to kill poultry is to use a chopping block in the backyard. Cut the chicken's head off and let it flop around until it is dead. It is, however, a very messy method, and you should have a box or basket to put over the bird to contain the blood.

A more popular way to kill poultry is by hanging the chicken by its legs at about shoulder height, or placing the bird head down in a killing cone. A killing cone is a cone or funnel made out of sheet metal in which you insert the bird, head hanging down out of the small end; the cone is attached to a wall and prevents breaking of bones and splattering of blood. Kill the bird by cutting its throat with a very sharp knife; the jugular vein can be cut from either inside the mouth or outside. Then insert the knife in the cleft of the roof of the mouth and cut a line between the eye and ear, twisting the knife. This is called debraining the bird, and its feathers are instantly loosened when this is done. Hang a weight through the lower beak to keep the bird from spraying blood, and catch the blood in a cup or can. A heavy wire soldered onto the can in the shape of an inverted **V** can be hooked into the lower beak to hold the mouth open.

Once the blood has drained out, you are ready to pluck the feathers. Plucking is easy unless the chickens are in molt. You may dry pluck, which leaves a very nice-looking carcass, or you can first scald the bird, an easier method which leaves an acceptable carcass. If you decide to dry pluck, you must debrain the bird to loosen the feathers.

To scald the chicken, you will need a kettle or pail large enough to dip the whole bird in. The temperature of the water is the most important factor for successful plucking; it should be between 126° and 130°F (52°C to 55°C). If the water is too cool the feathers will not loosen, and if it is too hot the skin will burn. Hold the bird by the feet, dip it into the hot water and move it around a bit. In about thirty seconds test to see if the feathers will pull out easily.

When the feathers are ready, hang the scalded bird at shoulder level and begin by picking off the wing and tail feathers with a strong twisting motion. If the feathers are hard to remove, return the bird to the hot water for a few seconds. The body feathers are pulled out next, and can be pulled out by the handful; then pluck the leg feathers.

Now only the pinfeathers (the small young feathers just emerging through the skin) are left, and these can be removed by pulling them out between a small knife and your thumb, or by using a strawberry huller. If you are plucking more than one bird, leave the pinfeathers until all the birds are ready for that point. Pinfeathers can be removed more easily from a cooled carcass.

Once the bird is completely plucked, cut off its head and feet. Leave a couple of inches of the neck and sever the feet at the joints. Once the carcass has cooled (refrigerate overnight or place in a freezer for two to three hours), proceed to clean the insides. Remove the intestines by cutting a line from the breastbone to the vent. Reach palm down into the body cavity under the breastbone and scoop out the intestines; the crop and windpipe require quite a bit of tugging. The chicken may then be cut into pieces and either cooked or frozen.

COSTS

The cost of raising chickens depends on how many and what type of chickens you are raising. You may not be able to compete with store-bought chicken on a price-per-pound basis, but the flavor and satisfaction will be worth the difference.

In our examples, we have used a unit of twenty-five Rhode Island Reds purchased as day-old straight-run chicks, and figured costs separately for broilers and layers. If you purchase all pullets, the price per chick will be about double. It costs approximately fifty-five

Annual Costs

Raising Broilers (*ten weeks*)	
Chicks (13@35¢ each)	$ 4.55
Feed (8 lb. per chick @ 10¢ per lb.)	10.40
Litter (50-lb. bag of shavings)	2.50
Total:	$17.45
Product Value	
Broiler meat (averaging 2.5 lb. dressed weight per bird – 32.5 lb. @ 85¢ per pound	$27.63
Profit:	$10.18
Raising and Maintaining Layers	
Chicks (12@35¢ each)	$ 4.20
Feed to 22 weeks (21 lb. per chick @ 10¢ per lb.)	25.20
Litter (50-lb. bag of shavings)	2.50
Feed for layers ($2\frac{1}{3}$ lb. per bird per week at $9.45 per 100 lb.)	85.05
Total:	$116.95
Product Value	
Eggs (240 eggs per hen = 240 dozen eggs @ 85¢ per dozen)	$204.00
Fowl (hens slaughtered at 18 months average 3.5 lb. dressed weight = 42 lbs. @ 40¢ per lb.)	16.80
Total:	$220.80
Profit:	$103.85

cents to produce a dozen eggs, including electricity for lights and refrigerator. We have used a price of eighty-five cents per dozen eggs, but many people are willing to pay more for fresh eggs.

CHICKS

For a fertilized egg to hatch it must be properly incubated. Using a hen to incubate is the easiest method, but an incubator is more reliable and economical. A hen ready to incubate is said to be broody. A broody hen is actually in a partial state of hibernation; her temperature drops slightly and her metabolic processes are somewhat slower. She will stay on the nest and defend it from other chickens and people. Provide a broody hen with a suitable nest, preferably on the floor where she will not be disturbed. A standard-sized hen, such as a Rhode Island Red, should have a nest about eighteen inches square with straw for nesting. When the nest is ready, make sure you dust her with louse powder before she sets; lice can kill chicks. Move the hen and two or three eggs to the nest at night. Once established on the nest, she will accept or lay more eggs. The eggs a broody hen incubates do not have to be her own.

Provide fresh water and food for the hen close to her nest. She will get off once or twice a day to eat and drink. The first eggs will probably hatch in twenty or twenty-one days, the rest of the eggs a day later. Use a slatted cover or a railing to make sure the hen does not leave the nest after the first chicks hatch.

Chicks and hen can leave the nest after the hatching is completed. If the weather is warm the chicks can be outside. The hen will need a small coop, about two-by-three feet, in which to brood her chicks. Pen the family to a small area for the first couple of weeks.

Some people use slats on one side of the coop to allow the chicks to get out while containing the hen for a couple of weeks. The advantage of confining the hen is that the chicks won't get tired, lost, or endangered while following her around. If you do this, do not forget to let the hen out for feed and water, or place them close enough for her to eat and drink through the slats.

An incubator provides heat, moisture, and air. The advantages of using an incubator are that you do not have to rely on a hen's going broody and you can control the incubation environment. The hatching rate for an incubator is about 70 percent. A hen may have a higher hatching rate if she sets full term, but there are many factors that may cause her to leave the nest, resulting in zero hatched eggs. An incubator is also more economical because the hens can continue their egg production undisturbed by hatching eggs and rearing chicks. There are two main types of incubators: still air and forced draft. New incubators cost from $30 to more than $100. To use an incubator, follow the manufacturer's instructions.

Raising chicks can be a lot of fun and rewarding, too. You will need some kind of brooder to provide heat. Brooders are available at poultry-supply stores, but homemade ones work very well and cost a lot less. The brooder can be kept in the barn or garage if you have electricity there, but in the house is probably safest. Wherever you keep the brooder, it must be safe from cats, dogs, and overenthusiastic children.

The simplest homemade brooder is a box with a partial cover and a light bulb for heat. Allow seven square inches per chick. Put news-

Homemade brooder

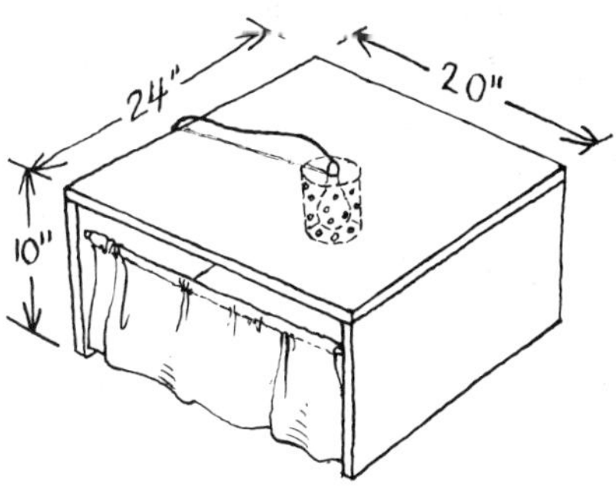

paper or litter in the bottom of the box and change it frequently. A partial cover helps eliminate drafts.

Suspend the light bulb far enough from the sides and floor of the brooder so the chicks do not bump into it. A spotlight reflector shield with clamp handle works very well.

You may also build a more permanent brooder that heats with a light bulb. If you use a brooder that lets the chicks come and go from the hover (the heated enclosure) you will need a cardboard or metal fence to keep the chicks near the heat source.

Ideal brooder temperature is around 90°F (32°C) for the first week, decreased by about 5°F every week thereafter. To make sure the temperature is right, just watch the chicks. If it is too cold they will all be huddled together under the bulb; if it is too hot they will be at the edges of the box, panting. Control the brooder temperature by raising or lowering the light bulb or changing its wattage.

The chicks will need the brooder for four to six weeks, by which time they will have most of their feathers. Let them become accustomed to the outdoors before then by putting them outside for a couple of hours and bring them back at night.

Install waterers and feeders in the brooder. A basin screwed on a quart jar makes an excellent waterer for a home brooder. Poultry-supply stores have chick feeders, but you can use a small can or jar lid.

It is very important to keep the litter clean and dry. Wet litter can chill the chicks and encourage the growth of coccidiosis.

MAINTENANCE

Debeaking is a method used to control cannibalism and discourage egg eating among chickens. Chickens may be debeaked at any age, but if you do it when they are day-old chicks it will need to be redone at about fourteen weeks of age. Using a sharp knife, make an incision about a quarter of an inch from the tip of the beak, then an upward cut toward the tip. Cut about halfway through, then break off the remainder of the beak.

You can also use an electric soldering iron or electric debeaker to burn the tip of the beak. This method has the advantage of being bloodless. Debeaked chickens cannot gather their feed from litter, and must be fed from feeders.

Chickens love to dust themselves, and if allowed to they will do so often. They lie on their sides and thrash around in the dirt. They especially love to dust in wood ashes. Dusting helps the chickens clean themselves and kills parasites. You should also dust the chickens periodically with an insecticide such as rotenone to help control mites and lice. There are also commercial products, usually containing Malathion, that can be sprayed around the chicken coop and over the roosts to control parasites.

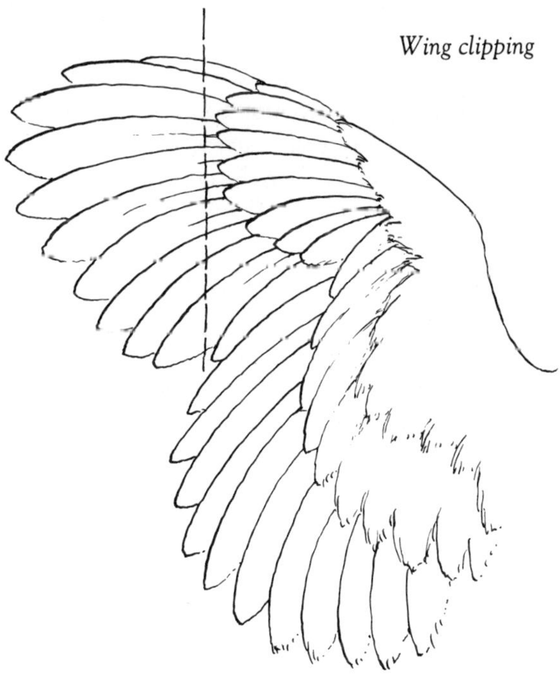

Wing clipping

A hen that is not laying is of no value to you and should be culled out of the flock and into the stew pot. A hen that is still laying has a bright red comb, old plumage which is molting, a soft abdomen, and a large, moist, white vent. The best test of a layer, however, is the distance between the bones on either side of the vent; in layers this distance is three or four fingers wide, whereas in nonlaying hens the distance is one or two fingers wide.

If you do not want your poultry to fly, clip the first ten feathers of one wing. The clipped feathers will eventually grow back, so reclip them as needed.

Chickens go into molt ten to fourteen months after they begin to lay eggs. Molt is a natural event designed to give the chickens rest. They start their molt in early autumn. They stop laying eggs and lose feathers. The molt lasts generally from six to ten weeks, at the end of which they will have regained their feathers and begin laying eggs again.

Not all the chickens in your flock will go into a natural molt at the same time, but you can use forced molt to speed up the process. You force a molt by creating a controlled-stress environment along with an absence or decrease of light. Keep the chickens in the coop, block out as much light as possible and feed little to no feed for two to three days (you must, however, keep fresh water available at all times). This will cause them all to go into a molt, at which time resume normal care.

Not all feather loss in chickens can be attributed to a full molt. Sometimes, due to minor stress, chickens may go into a partial molt of the neck feathers. Hens may lose back feathers due to roosters' activity. Feather loss can also be caused by parasites or fights with other chickens.

Cannibalism is a common problem that is frequently caused by crowded or improper living conditions, improper diet, or just plain boredom. If cannibalism develops, try to determine the cause and correct it.

Once the habit of cannibalism starts it is very hard to stop. You can put pine tar on the area of a chicken that has been attacked to create an objectionable taste, but debeaking your hens eliminates the problem.

Egg breaking is another negative habit; usually one hen will start but others soon learn from her. To correct this problem, remove the offending hen, make sure you are providing adequate nesting space, pick up eggs more frequently, darken the nest area, and put a glass egg in the nest; the latter makes egg breaking very unrewarding.

MEDICAL CONCERNS

There are several vaccinations that can be administered to chickens, but as a rule they should not be necessary for small flocks. If you are concerned about an immunization program, contact your county agent to see what, if any, fowl diseases in your area should be cause for concern.

A number of diseases and parasites affect chickens, but most can be prevented by buying from a poultry dealer with quality birds and by good management practices. If a bird should appear diseased, isolate it, observe it, and contact your county agent if an epidemic seems possible. Remember that occasionally a chicken will die and it does not mean an epidemic. To avoid medical problems provide adequate space, clean and dry quarters, a proper diet, and sufficient fresh water.

Coccidiosis is caused by a protozoan that cycles itself through soil or damp litter. The symptoms of coccidiosis are weight loss, sluggish activity and bloody droppings. To treat, add a commercial sulfur drug (available at poultry-supply stores) to the drinking water.

Crop bound occurs when the crop, which is used for storing food until it can be ground by the gizzard, becomes so packed with dry grass and food that the mass cannot leave it. A preventive measure is to make sure your chickens have sufficient grit and water. To treat a crop-bound bird, feed it mineral oil and water and gently massage the crop to work the solution around the mass.

Fowl pox is a virus. The early symptoms are dullness, loss of appetite, and a drop in egg production; advanced symptoms are brown or black scabs on the face and comb. There is no known treatment, and if the disease is prevalent in your area, young chickens should be vaccinated.

Frozen combs can occur when the weather gets too cold. A little petroleum jelly applied to the comb will keep the scab soft until the new tissue is healed underneath and the scab falls off.

Infectious bronchitis is a highly contagious viral disease. The symptoms are loss of appetite, coughing, sneezing, and sounds of respiratory distress. There may be a nasal discharge. There is no treatment, but the mortality rate for adult birds is negligible. A vaccine may be administered to chicks seven to ten days old.

Mites are exceedingly small parasites. There are red, feather, depluming, and scaly-leg mites. The most common mite is the red or roost mite. These parasites suck the blood of the bird they have infested, causing anemia and death. Commercial products are available to combat infestation. If one of your chickens has mites, assume they all do.

Newcastle disease (fowl pest) is also a viral disease. The symptoms are coughing, inactivity, and loss of appetite. Egg production ceases, and the last eggs laid are thin-shelled and off-color. There is no known treatment for the disease, but young chicks may be vaccinated against it.

Worms, or internal parasites, can affect chickens. The symptoms are weight loss, sluggish activity, stunted growth, and low egg production. There are worming medications, but it is difficult to determine what type of worm or worms are involved. Any time you butcher poultry, check for signs of internal parasites in the organs. If you suspect worms, call your county extension agent. To help prevent worms, maintain clean quarters, clean and disinfect the coop between flocks, and if you use a chicken yard, rotate the area every year.

Chickens belong on a homestead from an economic standpoint. The young, old, and everyone in between can care for them. They produce eggs and meat, can be shown at fairs, produce a cash product, and can be a seasonal or year-round project. In short, the chicken is truly a remarkable invention!

Chickens can be useful in the yard, as well as fun to watch. They eat insects, including wood ticks. Pet chickens can do a job on your garden, however, so you may need to fence them in when your garden is just starting. All in all, chickens are well worth the effort.

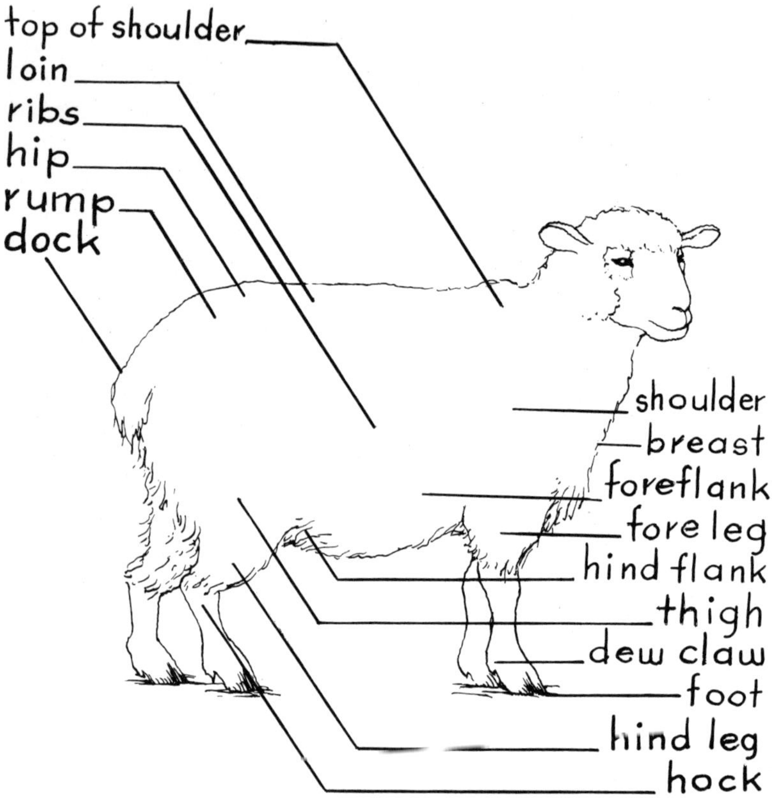

top of shoulder
loin
ribs
hip
rump
dock
shoulder
breast
foreflank
fore leg
hind flank
thigh
dew claw
foot
hind leg
hock

5

SHEEP

SMALL FLOCKS of sheep have become increasingly popular with homesteaders, who are interested in both the wool and the meat. Sheep are easy to care for and annually produce one or two lambs for the freezer and six to twelve pounds of wool per animal. Sheep are not expensive to raise, even without pasture, and their feed:meat conversion ratio is approximately 4:1. A lamb is ready to slaughter in six months or less, when it weighs 90 to 100 pounds, half of which is meat.

We got started in sheep when a ewe (a female sheep) was added to a truck-buying bargain. We wasted little time in finding an appropriate ram for her and we began to anticipate lamb dinners. We all enjoy lamb, but the cost of good cuts are virtually prohibitive for a family of five—all big eaters. In fact, the more we thought about it, the more ewes we decided to have. Our flock eventually numbered five ewes and one ram.

Shortly after completing our flock, we decided it would be fun for us to take our sheep for a short walk. The walk was absolute chaos as no two sheep went in the same direction at the same time. We all had a good time, but each of us greatly increased our respect for the shepherds who drive their flocks out to graze each day.

Sheep were domesticated thousands of years ago. Columbus brought the first sheep to the New World, and the Spaniards brought the Merino sheep to Florida in 1530. Until the Western ranges opened and transportation improved, sheep raising was done on a small scale.

The male of the species is a ram, the female is a ewe and the offspring is a lamb. A young castrated male is called a wether. An orphan is called a bummer, and a gummer is a sheep that has lost all of its front teeth due to age. A gummer is not a good buy, nor is a broken mouth, a sheep with some teeth missing because of advanced age.

There are three common ways of classifying sheep. The first is mutton type or a fine-wool type: the mutton type, such as the Lincoln and Romney, is raised primarily for meat; the fine-wool type, such as the Merino and Rambouillet, is raised primarily for wool. A mutton-wool type refers to the dual-purpose breeds, such as the Corriedale and Southdown.

A second method of classifying sheep is by length and coarseness of the wool: fine wooled, medium wooled, long wooled, and carpet wooled. The most popular breeds of sheep in this country are medium-wooled sheep, which are primarily meat animals with wool that is acceptable for spinning.

The third classification is to refer to sheep as either native or Western sheep. Native sheep are meat-type sheep and Westerns are fine wool or a cross of fine-wool and long-wool breeds. The fine-wool breeds are preferred on the ranges because of their herding instincts.

BREEDS

There are more than two hundred breeds of sheep in the world, but the ones that are commonly found on homesteads are the Cheviot, Columbia, Corriedale, Dorset, Hampshire, Oxford, Shropshire, Southdown, and Suffolk. The more unusual breeds of sheep may cause problems for the homesteader because of a lack of market, difficulty in obtaining a quality replacement ram or ewes, slow lamb development, or common lambing difficulties for the breed. There are crossbreeds that also do well for the homesteader, but be sure to choose quality stock and the right ram.

Cheviots originated in the Cheviot Hills of Scotland. They are a small breed, with rams weighing 160 to 220 pounds and ewes 120 to 160 pounds. They have no wool on their heads or below the knees and they are all white except for occasional black spots on the ears; they are polled, which means naturally lacking horns. There are two types of Cheviots, the Border or Southern and the North Country; the North Country is a much larger animal.

The Cheviots are excellent mothers and newborn lambs are exceptionally hardy. They often have twins and seldom experience lambing difficulties because the lambs have small heads. Cheviots tend to lamb late in the season. Cheviot wool is a good-quality medium wool that is easy to use in handspinning. The fleece weighs from six to eight pounds.

Columbia is an American breed developed in 1912 by crossbreeding Lincoln rams and Rambouillet ewes. They are large and hardy (rams, 200 to 275 pounds; ewes, 130 to 200 pounds) with no wool on the face below the forehead. They are all-white and polled.

Columbia ewes are good mothers, seldom having lambing difficulties, and produce lambs with an exceptional growth rate. They also have a calm temperament and are easy to handle. They have a medium wool that is excellent for handspinning. They produce a large fleece, generally more than ten pounds.

Corriedales originated in New Zealand about a century ago by crossbreeding Lincoln rams and Merino ewes, and were first brought to America in 1914. They are a medium-sized breed (rams, 185 to 250 pounds; ewes, 125 to 185 pounds), have facial wool, and are polled.

Corriedales are good mothers, having few lambing difficulties and good lamb growth. They are known for the longevity of their productive life. They have a high heat tolerance, which primarily affects the ram's resistance to sterility in hot weather but indicates that they are a good choice for the warmer sections of our country.

The Corriedale was bred as a dual-purpose breed, having both good wool and good meat production. The wool is medium and a great favorite of handspinners; the fleece often weighs more than twelve pounds.

Dorsets originated in England centuries ago. The traditional Dorset ram and ewe both have horns, but a polled strain was recently developed. Both strains are medium sized (rams, 175 to 200 pounds; ewes, 125 to 175 pounds), have some facial wool, and a relatively long body.

The ewes are prolific, often twinning, have good lambing ability and are excellent mothers. One of the pluses of the Dorset is an exceptionally long breeding season. This means that they can lamb in the late fall or early winter, providing lamb on the table when most breeds are just starting their lambing season. They are also known for a long productive life.

The wool of the Dorset provides about a seven-pound pure white fleece. The quality of the wool varies considerably, but if they are fed and cared for properly they can produce a satisfactory product.

Hampshires originated in Hampshire, England, and are well known for their heavyweight market lambs. It is a large breed (rams, 225 to 300 pounds; ewes, 150 to 225 pounds), with dark brown faces and lower legs, and some facial wool; they are polled.

The ewes are good mothers and produce a fast-growing, meaty lamb. The Hamp has a large head and broad shoulders, which can cause lambing difficulties. Homesteaders raising Hamps would be wise to use a Suffolk or other small-headed ram to make lambing easier.

The wool of the Hampshire is lightweight, usually six to eight pounds. The fleece frequently has black fibers in it, which detracts from the wool quality.

Oxfords originated in Oxford, England. They are a large hornless breed (rams, 250 to 350 pounds; ewes, 175 to 225 pounds), with medium gray or brown faces and legs. They have more facial wool than most Hamps and have a tuft of wool on the forehead which is longer than the rest of the facial wool. The fleece weighs ten to twelve pounds. The ewes are good mothers and seldom have trouble lambing. The lambs are fast-growing and meaty.

Shropshires, also English, are medium-sized (rams, 175 to 225 pounds; ewes, 130 to 175 pounds), with deep brown facial and leg wool. For awhile the Shrop was bred to increase the face wool, but the trend has been reversed because of the resulting tendency toward wool blindness. The ewes are excellent mothers, frequently twinning and seldom having lambing problems. The lambs are fast-growing.

The Shrop has a good-quality fleece, usually weighing seven to eight pounds; it tends not to have too many black fibers.

Southdowns, one of the oldest English breeds, is a small breed with a medium gray or brownish face and leg and a lot of short facial wool. They are considered the meatiest breed; the rams weigh from 160 to 225 pounds and the ewes from 125 to 160 pounds. The ewes are excellent mothers, but there can be lambing problems as the breed has a broad head. For this reason, both the Hamp and Southdown are not good breeds for the inexperienced homesteader. The Southdown fleece is fine with relatively short staples; it averages about six pounds.

Suffolk sheep originated in England and are the most popular breed in this country. They are a large polled breed (rams, 225 to 300 pounds; ewes, 160 to 225 pounds) with black faces and legs without wool. The ewes are good mothers, lambing problems are minimal, and the lambs are fast gainers. Their productive years, however, are relatively short. The Suffolk fleece has short wool and is lightweight, usually weighing four to five pounds.

SELECTING SHEEP

A prime factor in the type of breed you select is its availability in your area. The best time of year to buy sheep is the summer or early fall, and there is generally a larger supply of ewes. Prices also tend to be lower at that time of year.

It is best to start with just a few sheep and then grow to the optimum size for your needs. Your knowledge of sheep will increase and you can use your own best ewe lambs as additional stock when you decide to expand your operation. If you have fewer than four ewes, it probably would be economically sound not to keep your own ram.

You have a choice of purchasing purebred or grade sheep. A purebred will have the distinct characteristics of its breed and is eligible to be registered with its breed association. A grade sheep has the characteristics of a particular breed but is not eligible for registration. Purebreds cost more than grades and are not necessary for successful sheep raising. It is probably wiser for the noncommercial or inexperienced shepherd to start with grade ewes.

Choose ewes that are in a healthy, vigorous condition. Avoid animals that seem listless, lame, or not alert. They should have good conformation, which means a stocky body; clear eyes; a wide, deep chest; full spring of rib cage; wide back; wide and thick through the loin; straight legs; and large bones, giving a heavy, chunky overall look without being fat.

Check each ewe's mouth and do not accept any that have overshot jaws (upper and lower teeth that do not meet). Avoid gummers and sheep with broken mouths. Remember, however, that sheep do not have any top incisors.

It is possible to tell the approximate age of a sheep by its teeth. A lamb has eight small incisors. Two of the incisors per year are replaced by permanent teeth, so at age four the permanent teeth are in and the sheep is said to have a "full mouth." The permanent incisors are much larger and longer than their predecessors.

Determining age by teeth

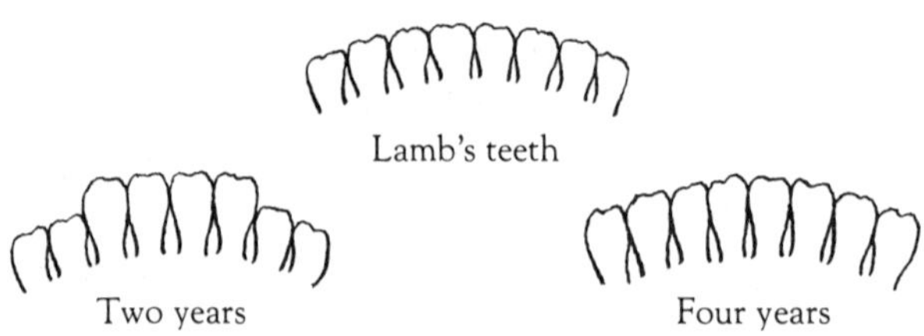

Approximately eight years

In addition to the eight incisors, adult sheep have twenty-four molars. After the permanent teeth are in they begin to slant forward, and with age the teeth spread out and wear down drastically. An old ewe's teeth look quite narrow.

Pay attention to how your perspective ewes walk, looking for ease of movement and lack of lameness. A good shepherd will have his sheep's feet trimmed. The ewe's udder should be soft, with two normal teats. Do not buy a ewe with teats that have been injured or are thick.

The fleece should be dense and have a nice luster. Avoid ewes with a ragged, uneven fleece, which may have resulted from excessive scratching due to ticks or mites. Part the fleece in several places and examine it closely.

In choosing a ram, remember that he will be half your flock: he will pass along his characteristics to every lamb he sires. Your ram should be especially strong in comformation where your ewes are weak. In general, he should be large, burly, and have plenty of bone. Do not choose a ram with a very large head because that characteristic can cause lambing problems. Check to make sure he has two well-developed testicles.

A good ram can handle up to forty ewes, and with proper care should be useful for about six years. You may use a ram under a year old if you have fewer than ten or twelve ewes. The most desirable age for a ram is about two years old. Purchase your ram a month or two before the breeding season begins, so that he will be acclimated before he needs to go to work.

The ram's personality needs to be mentioned: rams can be dangerous, so keep children away from them and be cautious yourself when in the pen with one. They ram with lightning speed and can cause serious damage.

You can control a ram by keeping one hand under his chin, which keeps him from getting his head down into butting position. We started controlling our ram by giving him a good back scratching. The trouble with that system is that any time we go into his area he expects us to scratch him, and if we do not, he threatens us.

SHELTER AND EQUIPMENT

A small flock of sheep can be successfully housed with other animals, but do not house them with pigs or cattle, as the risk of injury is too great. Goats and sheep are particularly good stable mates, both in terms of personality and equipment.

The primary considerations for a good sheep shelter are that it be dry, have good ventilation and no direct drafts. In warm climates, the sheep shelter need only provide shade and protection from heavy rains and the sun. Even in the coldest climate, sheep benefit from being closed in only during a winter storm or an early lambing.

Sheep need from sixteen to twenty square feet of floor space per head, the larger area for the larger breeds. A ram requires twenty to twenty-five square feet, a lamb eight square feet. Remember to allow for expected lambs when determining how many ewes you can handle.

Ideally the shelter should have a southern exposure, that it faces away from the wind and receives a maximum of light. Light is an important factor for both the sheep's health and your convenience. One square foot of window space for each twenty square feet of floor space is sufficient.

You may use either a trough or a combination grain and hay rack to feed your sheep. The trough should be at least eight inches wide and preferably flat-bottomed, so the sheep eat the grain slowly. Keep the trough clean. If you use a feed trough, you will still need a hayrack to ensure minimum waste.

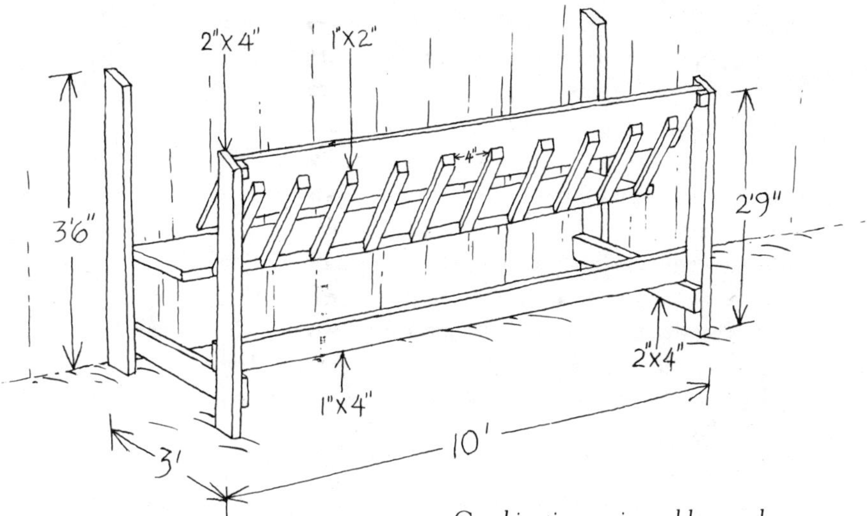

Combination grain and hay rack

A combination grain and hay rack is the most popular system with homesteaders. It may be designed to allow animals to feed on either side of the rack, or it can stand against a wall. A wide board at the top of the feeder helps to keep the chaff out of the animals' head and neck wool. Provide eighteen to twenty inches of space per ewe.

The watering equipment must be sufficiently large and easy to clean. The size of the waterer depends upon the number of sheep; one sheep averages $1\frac{1}{2}$ gallons of water per day. We have used large square galvanized tubs successfully.

By two weeks of age the lambs will begin to munch on hay, grass, and grain. To ensure that the lambs have access to feed and that the ewes do not overindulge, you need a creep feeder—a space large enough for lambs to get into, but small enough to keep sheep out. The openings to the creep should be eight to ten inches wide, and it should contain a waterer.

A lambing pen is necessary at lambing time to protect both the ewe and the lambs; they should stay in the pen for three days. An excellent lambing pen for those with limited space consists of hinged panels. If your lambing season occurs during cold weather, place a heat lamp over the lambing pen.

Our ewes all tend to lamb within a few weeks, so our lambing pen is converted to a creep feeder when the lambing is over. To accommodate this dual purpose, our pen has a two-by-four-inch board fastened vertically to both ends of the pen. The ends are then fastened to the walls of our barn to form a four-foot-square area. Instead of hinging the corner, we tie it together when it functions as a lambing pen and spread the corner to a ten-inch width when it converts to a creep feeder. We hammer in a board to maintain the opening and to keep the pen stable. When the lambs are gone, we remove the panels and store them until next year.

A good strong fence is a must to keep the sheep in and dogs out. The most common fencing is made out of medium-weight woven wire, 32 inches high with 6-inch stays. This is finished off with one strand of barbed wire on the bottom, running parallel with the con-

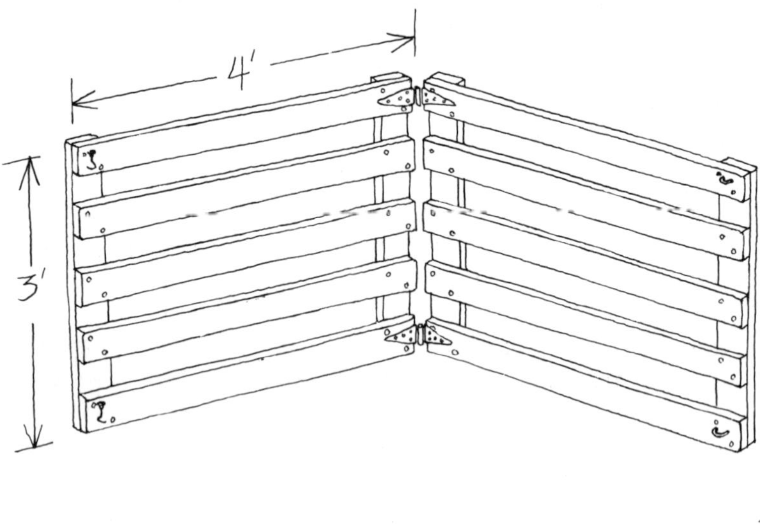

tour of the land, and two to five strands on top. The fencing is hung on $7\frac{1}{2}$-foot posts driven $2\frac{1}{2}$ feet into the ground and set 10 feet apart. Do not forget to treat your posts with a preservative before sinking them; coal-tar creosote is best, but used motor oil can be substituted for part or all of the creosote.

Barbed wire or boards may also be used for fencing. To be effective, barbed wire should be strung no wider apart than six inches for the first four rows, and then no wider than eight inches apart. The board fence need be no wider than four inches apart for the bottom three or four boards, and no wider than six inches at the top. String a row of barbed wire about two inches from the ground to discourage dogs from digging under the fence.

Standard electric fences are not very successful with sheep. They can be taught to respect the fence, but if they get excited they forget and can go crashing through the barrier.

If you plan to raise just a few sheep, the elaborate fencing and pasture is hard to justify in terms of land usage and expense. For most homesteaders, the feedlot system is the most practical. Instead of pasture and fencing, you use only the shelter and an exercise yard, feeding the sheep hay and grain exclusively.

Ideally, the exercise yard is an extension of the shelter area and has a good slope for drainage. Allow approximately thirty square feet of space per head. You need good fencing for this area, but it can be done at a reasonable cost, compared to fencing in an acre or so of pasture.

BREEDING

For most breeds, the breeding season is from August or September to January. The Dorset can breed in the spring and lamb "out of season" in the fall. The gestation period is 145 to 155 days.

A ewe is in heat an average of thirty-three hours, with the interval between heats about sixteen days. Ovulation takes place near the end of heat. You may have a ewe bred when she is about nine months old if she is in good condition, but chances of lambing problems are greater for a ewe under a year old.

As the shorter and cooler days of the breeding season begin, you can control your lambing season to fit your setup by removing the ram from your ewes until you are ready for them to be bred. If you would prefer to avoid a January or early February lambing because of severe weather, separate your ram from the ewes until October or November.

Flushing—increasing a ewe's feed so she gains weight—is recommended to increase the number of lambs produced and decrease the time it takes for a ewe to settle (conceive). U.S. Department of Agriculture studies indicate that the increase in number of lambs can be as much as 18 percent for two or three weeks of flushing before and after breeding takes place.

To start flushing, feed $\frac{1}{4}$ to $\frac{1}{3}$ pound of a grain ration; in a week, gradually work each ewe up to $\frac{1}{2}$ or $\frac{3}{4}$ pound of grain per day. Two weeks after breeding, start to decrease the grain ration, then stop feeding it three weeks after breeding took place. Flushing can also be accomplished by providing abundant lush pasture.

Flush your ram to ensure that he is in the peak of condition and performs well. If you do not have a lush pasture, your ram should be given about one pound of grain per day through the breeding season.

Before you flush your ewes, deworm your sheep and tag out and crotch them. To tag out means to clip away the wool around the vulva below the dock (tail) to aid the ram in servicing her. Crotching is clipping the wool from the udder and inside the flanks. If you do so at this time, you should not have to do it when she is well into the gestation period, for crotching makes lambing easier.

It is difficult to tell when a ewe is in heat and therefore difficult to tell when she has been bred. You may purchase a marking harness for the ram to wear. It fits over his shoulders, around his chest and between his front legs. The harness is equipped with a holder for a marking crayon; when he mounts the ewe, the crayon leaves a mark on her. Change colors after two weeks to make sure the ewes have settled by an absence of new markings. The harnesses are a bit expensive ($8 to $15), so you may use just a marking crayon or chalk and liberally mark the ram's lower chest.

Since ewes start to make bag (show udder development) about one to two weeks before lambing, you may also check for bred ewes once a week starting in January. The only problem with this method is that an infertile ram would not be detected until too late in the season for a lamb crop.

There are four breeding methods that can be used. One is inbreeding, the breeding of close relatives (father-daughter, sister-brother). This method is not recommended for sheep because negative characteristics will be perpetuated in the offspring and no improvement can be made in your flock. Linebreeding is the mating of less closely related animals, such as aunt-nephew, with the idea of retaining a close relationship to some excellent ancestor. Outcrossing is the mating of unrelated animals within the same breed. Outcrossing can be highly successful if you know that the line consistently produces characteristics that you are trying to breed into your line. Crossbreeding is the mating of two different breeds, such as a Dorset ewe and a Shropshire ram.

The best breeding methods to use with sheep are crossbreeding, outcrossing, and linebreeding. Inbreeding only becomes a problem if you keep the best of your lambs for breeding stock. When they are ready to be bred, replace your ram.

LAMBING

The ewe's gestation period can be divided into two parts; the first three months and the last two months. Each period requires different care and feeding practices.

During the first twelve weeks you want the ewe to gain a little weight, but you do not want her to become fat. It is difficult to tell visually how much weight a ewe is gaining, so get in the habit of feeling the ewe, especially along the backbone and ribs, for indications of weight increase. During the first part of pregnancy the ewe can be adequately maintained on regular rations for a dry ewe.

During the last six weeks of pregnancy, approximately 70 percent of fetal growth takes place and the ewe's nutritional needs alter accordingly. Your goal is to have slow but steady weight gain. A suggested ration is $\frac{1}{4}$ pound of grain per day for the first week, $\frac{1}{2}$ pound the second week, and then gradually increasing to 1 pound per day as long as the ewe is not gaining excessive weight. The grain may be any good mixed dairy feed containing 12 to 14 percent protein, or 16 to 18 percent protein if your hay is not a high-quality legume hay. The ewe should still be offered 3 to 4 pounds of hay per day once you start graining. Variations in the diet depend upon the ewe's general condition, the breed of sheep, and individual differences.

You may also want to feed the ewe molasses during this period; use livestock molasses, which is much cheaper than the commercial kind. About a pint per day mixed in with the water (or $\frac{1}{2}$ pound of dry molasses) is a precaution against lambing paralysis.

After lambing, do not feed the ewe more than $\frac{1}{2}$ pound of grain per day for two or three days, although she should have all the roughage she wants. On the fourth day, continue her roughage and start adding some more grain. At about a week after lambing, increase the grain so that by ten days after lambing her grain rations are up to about 2 pounds per day. As the lambs approach weaning time, begin to taper off the grain ration so that by weaning the ewe is just eating hay. This will help the ewe properly dry off her milk production.

Handle the pregnant ewe carefully to avoid problems; rough handling or undue alarm can cause abortion.

Check your previous tagging and crotching jobs to make sure the areas are still trimmed; if not, trim them. If your ewes have very wooly faces, it is a good idea to carefully trim the wool away from the eyes to aid the ewes in keeping track of the lambs.

As the end of gestation approaches, set your lambing pen up, using lots of clean hay or straw for bedding, and watch for signs of ewes' preparing for delivery. The first sign is making bag. Other signs that delivery is close at hand are the ewe staying off by herself, a general restlessness with much pawing of the ground, turning around repeatedly lying down and getting up. Do not count on seeing these symptoms, however; one time we had a ewe drop a lamb while running with the flock to the feed trough!

Successful lambing is not due to chance, but to the conscientiousness of the shepherd. Once a ewe starts to make bag, check her several times a day for any unusual behavior. If you think a ewe might be in labor, put her in the lambing pen and check on her every couple of hours.

Have a lambing kit ready; this includes clean towels, tincture of iodine and a small cup, lubricating ointment, a pessary, and a good antiseptic soap or a pair of sterile surgical gloves.

Once the ewe starts to pass strings of mucus, which will be pink or mixed with blood, you know that the lamb or lambs are on the way. A ewe can have a single birth or twins, and even occasionally triplets. She will grunt and strain and in fifteen minutes or so the water bag will appear. It will look like a dark bulge. Next you should see two feet and then the lamb's nose. The ewe will work hard for a few more minutes, until the head and shoulders have been delivered. The rest of the lamb comes out easily.

If things are going well, do not interfere. Sometimes lambs are delivered hind legs first and then you will first see the bottoms of the back feet. This is also a normal delivery position, but the delivery must be relatively fast. Watch for the breaking of the umbilical cord; once it is broken the lamb starts to breath, so you want it out of the ewe as quickly as possible. When the lamb is out to its shoulders, pull firmly downward as the ewe contracts. Do not jerk the lamb; provide a steady, firm, downward pressure. Work with the ewe.

When the lamb is born, the ewe will lick off the amniotic fluid. If the ewe is exhausted or busy delivering a second lamb, you may rub the lamb down, paying special attention to cleaning the mucus away from its nose and mouth. If it is extremely cold, turn the heat lamp on. If the lamb is not breathing or sounds bubbly when it is born, hold it upside down by its back legs and slap its side firmly with your open hand to force out the rest of the mucus and fluid.

If a ewe is in labor forty-five minutes or more without producing a lamb, check on the position of the fetus. First, scrub up with an antiseptic soap, dry your hands and lubricate them or put on sterile gloves. The procedure is much easier if you have an assistant holding the ewe while you do the checking.

If you can see or feel the two front legs but continued contractions do not produce the head, the head is probably turned back. Carefully move your hand up one of the legs to the chest and make sure that both legs belong to the same lamb. Move up the neck and see where the head is located. Gently draw the head into position on top of the legs. You may have to push the front legs back a bit to get the head into position. Occasionally one of the legs will be bent under at the knee. If this is the case, push the lamb back and then gently move the bent leg into proper position. Once the lamb is in position, you may gently assist the ewe by pulling downward on the lamb's legs when the ewe contracts.

Another abnormal position is the fetus's coming head first and upside down. Gently push the fetus back and rotate it right side up, then gently extend the front legs into normal position.

If nothing shows for the ewe's efforts but the water bag, the lamb may be breeched, meaning the buttocks are being presented first. Push the fetus forward and bring the hind legs into position for a normal back-end delivery position.

If you find that there appears to be more than one lamb trying to be born at once, you will have to separate them. The problem is to match appendages with the proper lamb. Usually the lamb whose head is in the birth canal is the one you want to be born first; the twin should be pushed back to await its turn. Take your time and move gently and carefully. Make sure you know which legs belong to which lamb before you push or pull. Once you have separated them, gently slide the first lamb enough ahead so the second one does not interfere with its birth.

After assisting with a birth, and once the afterbirth has been passed, insert a pessary to guard against infection. If your efforts to assist the ewe are not successful within ten or fifteen minutes, call the veterinarian.

When the lambing is over, treat the umbilical cord. Fill the cup with tincture of iodine, hold the lamb so the cord is in the center of the cup, press the cup against the lamb and tip the lamb up so that the whole area is treated.

The lambs will want to nurse within an hour after birth. Make sure that they are actually getting milk and not just going through the motions: if the lamb is nursing, its mouth will feel warm to your finger and you should be able to detect its throat muscles contracting as it swallows. If the lamb is not getting any milk, squeeze the ewe's teats until you get a squirt of milk from each one; this process removes the wax plug that forms.

Lambs should have their tails docked when they are from one to two weeks old. Ram lambs not intended for breeding should be castrated at the same time the tails are docked. These procedures are described under "Maintenance," pages 133–135.

When the ewe is through lambing, offer her a bucket of warm water to drink; this helps replace some of the heat lost while lambing.

Once the ewe has passed the placenta, you may either bury it or let her eat it. Make sure that the soiled and damp bedding is replaced with clean, dry bedding.

Problems

Lambing paralysis can result from the inability of the liver to transform body fats into sugar or from an inability of other tissues to meet the carbohydrate demands of the ewe and the unborn lamb or lambs. It seems to affect ewes that are too fat or malnourished. Symptoms include the ewe's acting listless and weak, twitching muscles, aimless walking, or coma. Call your veterinarian if you suspect this problem; it can be fatal.

Milk fever can occur from six weeks before to ten weeks after lambing. The condition is characterized by calcium being used up in milk production faster than the ewe's glands can replace it. The symptoms are a weakness in the rump and back legs and a subnormal temperature; she may drag her back legs, will not be alert and her eyes will look glazed. Call the veterinarian if you suspect milk fever.

Occasionally a ewe will not pass the placenta. Do not attempt to remove it yourself. If the ewe has not passed it after twelve hours, call your veterinarian.

A prolapsed uterus is caused by the ewe's continuing to strain after the lambs are born and thus forcing the uterus out. It most frequently happens after a difficult birth. You may call the veterinarian to try to replace it, but you are probably wiser to slaughter the ewe. Once a ewe has prolapsed her uterus during lambing she is likely to do so again at the next lambing. Prolapsing tends to be an inheritable trait, so you would be wise not to retain a ewe lamb from the ewe for future breeding stock.

Mastitis is an inflammation of the mammary glands due to the presence of bacteria. The symptom of mastitis in sheep is a cold, bluish udder and/or teats. If this problem develops, call your veterinarian immediately.

CARING FOR LAMBS

By the time the lambs are two weeks old they should be offered both grain and hay in the creep feeder. You may also buy complete lamb feeds, usually pelleted, which eliminate the need for hay and other grains. The creep trough should be low and shallow, with a board lengthwise above the trough to keep the lambs from jumping into it. Provide all the fresh feed they want twice a day, and feed any refuse to the ewes. Feed the hay after the grain, preferably from a rack.

Any dairy grain of 14 percent protein content may be fed, or you may use a combination of grains, such as half corn, half oats. Feed a bulky grain, such as oats, for the first few days and then gradually increase the amount of other feed. The lambs will consume very little at first.

If you have pasture for your sheep, get both ewes and lambs gradually accustomed to it by letting them on pasture for only two to four hours at first. They need to continue their regular feed as supplement until they are adjusted to the pasture.

Plenty of salt and fresh water should always be available for the lambs; provide them in the creep.

Lambs may be weaned at 3 to $3\frac{1}{2}$ months if they are grain fed and at about 4 to 5 months if pasture fed. Ideally, the lamb and ewe should be physically separated and out of hearing range of each other when weaned. The lambs are ready for slaughter at 90 to 100 pounds, which should be reached by 4 to 6 months.

SUGGESTED DAILY RATIONS FOR LAMBS

Month	*Hay*	*Grain*
1	free feed	free feed
2	free feed	free feed
3	$\frac{1}{4}$ lb.	$\frac{1}{4}$ lb.
4	$\frac{1}{2}$ lb.	$\frac{1}{2}$ lb.
5	$\frac{3}{4}$ lb.	$\frac{3}{4}$ lb.
6	1 lb.	1 lb.

It is not uncommon for a ewe to reject her lamb. Sometimes you can fool the ewe by rubbing vanilla on her nose and on the lamb's rump. If she still does not let the lamb nurse, you can force her to let it do so: hold her back legs and have someone else hold her shoulders and let the lamb nurse. This system sometimes works with an orphan lamb and a foster mother.

If a ewe completely rejects a lamb or if a lamb is orphaned, you will have to raise it. Keep the lamb in a warm, dry place separate from the ewes, and feed it every four hours for the first week. It will take only one to two ounces at each feeding at first, but gradually the amount may be worked up to ten ounces per feeding, three times a day. The milk should be warmed to 103°F (40°C); you may use a baby bottle or a lamb nipple that fits on a soda bottle.

The first milk from a ewe is called colostrum, and it contains antibodies that protect the lamb against some diseases; if at all possible, the lamb should have it. You would be wise to have some frozen colostrum on hand just in case you lose a ewe while lambing. A substitute of colostrum formula can be made with eight ounces of milk, a beaten egg yolk, $\frac{1}{2}$ teaspoon of sugar, and $\frac{1}{2}$ teaspoon of cod-liver oil. Feed the colostrum or the formula for two to three days.

Lambs do well on goat's milk or lamb-milk replacer. Cow's milk does not have sufficient fat or protein for the lamb to thrive.

You may wean a bummer (orphan) at thirty to thirty-five days or twenty-five to thirty pounds, if you have the lamb eating a concentrate such as calf manna along with the normal grain and hay ration. Teach the lamb to eat calf manna early by holding about a tablespoon of it in your hand and letting the lamb suck your fingers, allowing the pellets to roll into its mouth. As the lamb eats more calf manna, gradually dilute the milk with warm water. Continue the bottle until the lamb is consuming $\frac{1}{4}$ pound of the concentrate per day.

FEEDING

The diet for sheep is relatively uncomplicated. High-quality pasture or legume hay can be the total diet, except for just prior to and after lambing. Most homesteaders do not have a great deal of pasture for sheep, and therefore their sheep are usually raised on hay and grain or a complete pelleted feed.

The best pasture for sheep is a mixture of alfalfa or clover and grass. These combinations reduce the danger of bloat. But you must gradually introduce your flock to lush pasture and guard against turning them out on pasture on hot days after a rain or while the dew is still present.

To maximize the benefits of pasture, plan to use a rotation system. Fence the pasture into sections and put your sheep in a new section before they overgraze the previous pasture. The best pasture is kept short, not more than three to four inches high. The system of rotation grazing also helps control parasitic disease.

How many sheep can be fed on an acre of land depends upon the quality of your pasture, which is determined by the type of vegetation, type of soil, amount of rocks, and rainfall. The best-quality pasture, in all respects, can handle up to fifteen ewes plus lambs per acre; poor to average pasture can handle two to seven sheep, respectively. The best method for determining how many sheep your particular pasture can handle is to start with just a few sheep and see how it holds up. You must supplement pasture with grain or legume hay if your pasture does not meet the sheep's nutritional requirements both in quantity and quality.

Sheep may be fed entirely on high-quality hay, with the exceptions of the six to eight weeks before lambing, the nursing period, and when flushing. Good-quality legume hay, or mixed legume and grass hay containing at least 50 percent legumes, may be fed in rations of 3 to $4\frac{1}{2}$ pounds of hay per head per day as a complete diet. If the only hay you have is good-quality grass hay, you will need to supplement it with a protein supplement or grain.

The easiest grain to feed with grass hay is a mixed dairy feed or horse feed. You may also use other grains, such as corn, barley, wheat, and oats. Grain for sheep does not have to be ground.

SUGGESTED DAILY RATIONS FOR DRY EWES
UP TO SIX WEEKS BEFORE LAMBING

Ration	*Roughage*	*Grain*
1	Pasture, free feed	none
2	Legume hay, 3 to 4 lb.	none
3	Timothy, 3 lb.	14% protein dairy feed, 1 lb. (or $\frac{1}{2}$ lb. protein supplement)
4	Timothy, 3 lb.	Corn, oats, wheat, bran, $\frac{1}{2}$ lb. each.
5	Grass hay, 3 to 4 lb.	16 to 18% diary feed, 1 lb.

SUGGESTED DAILY RATIONS FOR EWES
FOUR TO SIX WEEKS BEFORE LAMBING

Ration	*Roughage*	*Grain*
1	Legume hay, 3 to 4 lb.	Oats, corn, bran, $\frac{3}{4}$ lb.
2	Legume-grass hay 3 to 4 lb.	Oats, corn, or 16 to 18% protein dairy feed, 1 lb.

If you do not have pasture land or much storage area for hay, the cheapest diet may be a complete pelleted feed. A complete feed for sheep may be difficult, if not impossible, to come by in your area. If this is the case, use a complete feed for horses. You may feed grass hay to dry ewes and your ram while the supply and price are right and then gradually switch to the complete feed during the winter.

Salt should be fed free choice in a small box. You may wish to add one part phenothiazine to ten parts salt to aid in the control of parasites, if your sheep are on pasture.

If you are not using a complete feed pellet, you may also wish to add calcium and phosphorus to your sheep's diet in the form of bone meal, or its equivalent, fed free choice.

Sheep do not like to drink water that is not clean, so provide plenty of clean, fresh water. Sheep will consume about 1 to $1\frac{1}{2}$ gallons of water per day; lactating ewes possibly need a bit more.

COSTS

The cost of raising sheep varies with the amount of pasture available. Most homesteaders do not have sufficient quality pasture available, and so the estimated costs presented do not allow for this free feed. Obviously, the costs will be significantly lower—about half—if you use pasture. The costs will also be lowered if you supplement with feed from the garden.

We have used a unit of four ewes and one ram and have assumed a 150 percent lamb crop. The figures are based on a medium-sized breed fed according to the feeding tables in this chapter. We have not included the original price per head for the breeding stock, as this varies significantly according to age and breeding. You should include the original prices in determining your actual costs, but allow for this price to be spread over the number of expected years of productivity. If you do not own a ram you would, of course, eliminate the ram's feed costs.

Annual Costs

Hay (Legume)		***Feeding Rate***			
1 Ram, 4 Ewes		3.5 lb./day per sheep × 365 days			
		6,387 lb.			
6 Lambs (assume all born in same month; 4 sold and 2 in freezer at end of 6 months)	Month:	1	—	4	90 lb.
		2	—	5	135 lb.
		3	45 lb.	6	180 lb.
			450 lb.		
Total Hay Cost:	6,837 lb. @ $80 per ton = $273.60				
Grain (Corn/Oats)					
1 Ram: 1 lb. per day for 50 days (breeding season)			50 lb.		
4 Ewes: 6 weeks prior to lambing and lactating (assumes gradual increase and decrease of grain rations)			630 lb.		
			680 lb.		
6 Lambs	Month:	1	—	4	90 lb.
		2	—	5	135 lb.
		3	45 lb.	6	180 lb.
			450 lb.		
Total Grain Cost:	1,130 lb. @ $9 per CWT = $101.70				
					273.60
			Total Feed Cost:		$375.30

Income

Sale of lambs: 4 lambs (90 lb. each) @ $.60 per lb.	$216.00
Sale of wool (excluding lambs): @ $3.50 per fleece	17.50
	$233.50
Shearing cost (@ $2 per sheep)	− 10.00
Total Income:	223.50
Profit:	$151.80

2 remaining lambs (90 lb. each) dressing out at 50 percent, provide 90 lb. of meat @ $1.69 per lb.

The cost per pound of meat may seem quite high compared to other stock found on a homestead, but keep in mind the cost of prime lamb at the butcher shop.

SHEEP

MAINTENANCE

Sheep are sheared once a year, usually after the weather has warmed up enough to bring out the grease in the wool, making it easier to cut. Some people shear their ewes before they lamb but we do not, because it is still apt to be cold and we prefer not to risk any injury to a pregnant ewe.

When it was time to shear our first sheep, we had a pair of hand shears, but no instructions. We looked through every book available at the time and the only reference we found to hand shearing was in one of the *Foxfire* books, which said "nobody hand shears anymore." So much for that. Since then we have learned that you use the same motions for hand shearing as you use with electrical shears. Start with the head and work back to the tail, keeping the fleece all in one piece.

Shearing is not difficult and can be done in a few minutes, if you know what to do and how to do it. Before it is time to shear, contact a sheep-raising neighbor or your county extension agent and find out who does shearing in your area. Their services usually run $2 a sheep. After observing them, you may decide whether or not you want to add sheep shearing to your lists of talents.

One of our greatest pleasures from homesteading is the wool blankets we have had made from the wool from our sheep. Each year we send our wool, in the grease (as it comes from the sheep), to the Shippenberg Woolen Mills in Shippenberg, Pennsylvania. We stuff the fleece into feed sacks, tag them, and lug them down to the local post office with great pride. In about six weeks we get back 100 percent wool blankets, in colors we have chosen, for a fraction of the price we would pay at a store.

Sheep dipping refers to the process of applying an insecticide to the sheep after shearing to destroy sheep ked or sheep ticks. This parasitic insect is actually a fly and is the most common external parasite of sheep. The large dipping vat, from which the process gets its name, is obviously not practical for a few sheep, but sprinkling or dusting on an insecticide can be effective. Government regulations on insecticides change so quickly that you need to check with your extension service or farm-supply store each year for what is currently available.

Healthy ram lambs should be castrated between seven and ten days old. It is advisable to castrate and dock at the same setting, but do the castration first to avoid bumping the dock.

Have a helper hold the lamb upside down on his or her lap or on a table, holding a front and hind leg in each hand. Force the testicles up against the bottom of the scrotum and cut away the bottom third of the scrotum with a sharp knife or razor blade. Gently squeeze the base of the scrotum to force the testicles and tunic (the white membrane around the testicles) out. Keeping one hand at the base of the scrotum to keep the testicles from going back in, pull one testicle with your other hand until it breaks; repeat with the other testicle. When finished, pour tincture of iodine over the area. If the fly season has started, apply a fly repellent, such as pine-tar oil.

Lamb's tails should also be docked at seven to ten days to control maggots and to help the ewes receive the ram more easily. The best method of docking is to use an emasculator. Place the emasculator curved end down on the tail to give the lamb a one-inch stub. Apply steady pressure until the tail drops off and then continue pressure for thirty seconds more. The additional pressure is applied to make sure all the blood vessels are crushed, resulting in a bloodless dock.

You may also dock with a sharp knife. Pull the skin of the tail toward the lamb's body and cut off the tail, leaving a one-inch stub. This method causes bleeding; if excessive bleeding lasts for more than a few minutes, tie a string around the stub. Apply a coagulant (or alum), then remove the string after four hours. Dip the dock into some tincture of iodine when you are finished. If the fly season has started, use a fly repellent.

Sheep hoofs should be trimmed at least twice a year to ensure a flat surface on the bottom of the hoofs, which helps prevent lameness.

The sheep needs to be sitting on its rump, like a dog begging, for you to do the job easily. To get the sheep in this position, stand on the right side of the animal by its head and slip your left thumb into its mouth, in back of the incisor teeth, and place your right hand on the sheep's left hip. Bend the sheep's head over its right shoulder as you press your hand down and bring the sheep toward you. Pull the upper part of the sheep's body into a sitting position and you are ready.

Using a jackknife, pruning shears, or a hoof knife, clean out any dirt or manure that has accumulated. Trim off any portion of the hoof that is folded over, then trim the remainder of the excess hoof. You want the horn to be level with the sole and not too extended in front. A look at a lamb's hoof will show you the ideal shape.

MEDICAL CONCERNS

The best methods for keeping sheep healthy are proper feeding, management, and sanitation. Symptoms indicating a sick sheep include a loss of appetite, weakness or staggering, labored or fast breathing, and an above-normal temperature; normal sheep temperature is 102°F (39°C). There are some common diseases with which every shepherd should be familiar.

Enterotoxemia, also known as overeating disease, is caused by an intestinal bacteria. It is usually seen in animals that have been overfed on lush pasture or heavy grain rations. The symptoms are signs of colic, with frequent getting up and down and general restlessness. Advanced symptoms are unsteadiness and convulsions. Frequently, affected lambs will die within twenty-four hours. Treatment is generally not successful, but this disease may be prevented by an appropriate feeding program or by an enterotoxemia vaccination. This may be given annually to mature sheep and to weanlings.

External parasites, such as the sheep tick or ked and lice, will be eliminated by using an insecticide after shearing and treating your lambs.

Sheep scabies are caused by minute gray mites that burrow under the skin. The skin becomes inflamed and a gray scaly crust forms; the wool in that area falls out. This problem is difficult to eradicate and so contagious that you should immediately contact your veterinarian once it is detected.

Wool maggots result from flies' depositing their eggs on cuts or on dirty wool. The eggs hatch into maggots, which live off the sheep's flesh. It is relatively easy to get rid of them: locate the area of infestation, clip all the wool away from it and treat with commercial products for this problem. Tail docking and tagging your sheep help to reduce the incidence of maggots.

Foot rot is a contagious bacterial infection to which sheep of all ages are susceptible. The first symptom is usually lameness, and the infected foot will have a very bad odor. To treat this disease, you must first trim all of the overgrown part of each hoof, including all ragged or separated pieces of the horn and sole. Then pare all of the diseased tissue until you have removed every pocket and crack. Check carefully for every hidden or deep pocket of infection. You may have to cut healthy tissue away to get at an infected area.

When the trimming is finished, liberally brush on a commercial treatment product according to the directions. Then have the animal stand in the solution for at least four minutes. Soak each infected foot every other day for two weeks and retrim hoofs of infected sheep as necessary.

To decrease the chances of foot rot, keep the hoofs of your sheep trimmed at all times and do not keep your sheep on wet ground. If they are frequently on wet ground, the skin gets irritated and often injured by wet material packed between the claws, which makes sheep susceptible to any foot-rot bacteria.

Internal parasites are sheep's biggest health problem. Sheep on pasture are particularly hard hit because they graze close to the ground and ingest worm eggs, small worms, and snails that carry parasites. You will not be able to rid your sheep of all internal parasites, but you can control the level of infestation. Give a broad-spectrum wormer in the spring and fall. Once your flock is established, have a veterinarian examine a fecal sample to make sure your sheep are not harboring any parasites that a broad-spectrum wormer will not kill.

Pneumonia is caused by bacteria. A sheep with pneumonia will often stand with its head down, show labored breathing and will tend to run a slightly elevated temperature. If you suspect this disease, isolate the animal and call your veterinarian.

Tetanus or lockjaw can affect sheep. Keep your sheep area free of objects that can cause wounds and treat any wounds by washing them thoroughly with soapy water. Once the wound is clean, use iodine or another antiseptic solution. Tail docking and castration should be checked until thoroughly healed to make sure you are not inviting tetanus. You may vaccinate against tetanus.

Without ample pasture, sheep are not as practical a meat animal as others you might raise. If you appreciate the delicacy of the meat, however, and if you maximize the wool products, they can be justified on your homestead.

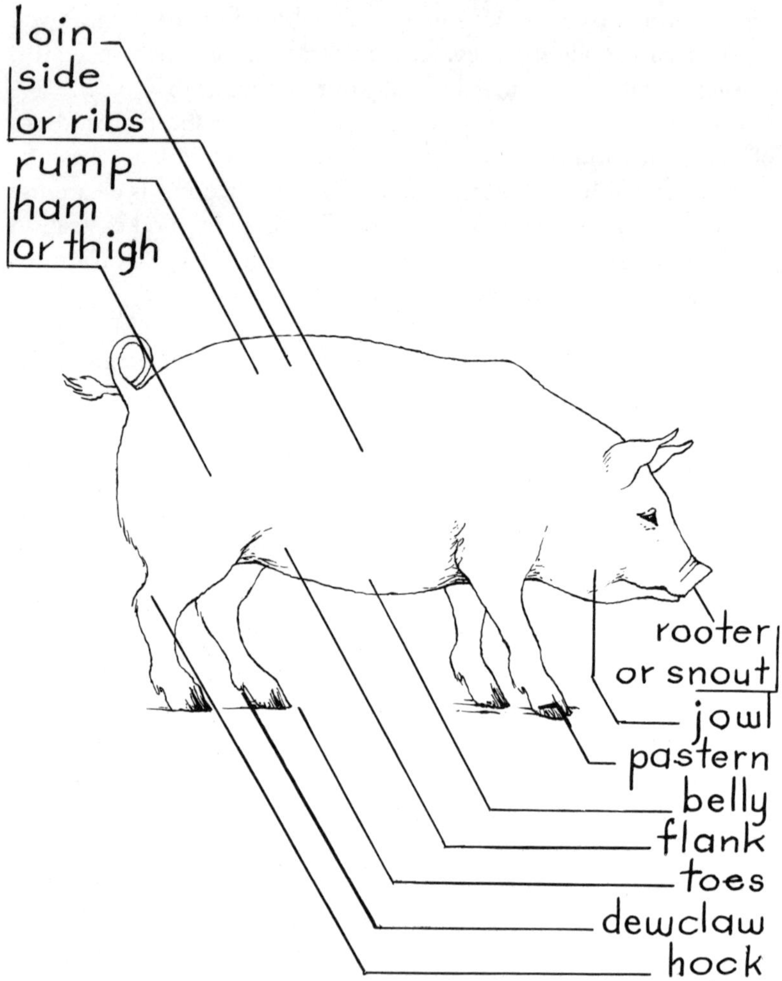
loin
side
or ribs
rump
ham
or thigh
rooter
or snout
jowl
pastern
belly
flank
toes
dewclaw
hock

6

PIGS

PIGS ON A HOMESTEAD are a natural. Their ability to convert kitchen and garden waste into delicious meat is well known. Not only will the meat be cheaper per pound than at the local grocery store, but you will also find that the meat has more flavor and tenderness than can ever be obtained by mass production. Figure that a 225-pound carcass will give you 150 pounds of meat cuts and about 25 pounds of excess fat. Pigs have a feed:meat conversion ratio of 6:1.

Side products from raising pigs are also a plus; for example, you can make nine pounds of soap for about 75¢. You will also have some of the finest manure you can get for your garden. If you work at it, nothing but the squeal need go to waste.

An additional bonus with raising pigs is their personalities. They are very intelligent, and as a result, a lot of fun; in fact, quite a few pigs have become family pets. However, if having a pet pig is your goal, you would be better off getting one of the miniature pig breeds used for laboratory research, as their mature weight is only around 225 pounds, as opposed to 600 to 1,000 pounds for standard-sized pigs.

Some of our funniest memories are of trying to round up pigs that have gotten loose. Ricky and I once chased our first pair of pigs through the cornfield for almost an hour. Since Dick was safely in Boston I called Nappy, the guru of homesteaders in our area, to see what we should do. He promptly put me in my place by saying "it is not what you should do, but what you should be, which is smarter than the pig." When Dick got home he poured some milk into a pail and the pigs just followed him back into their pen.

A friend of ours had a rough couple of weeks with her pigs during a hot spell. No matter what she tried, by midmorning people would stop by to tell her that her pigs were in a nearby creek. Among the characteristics of pigs, one must include extreme stubbornness; trying to move three pigs out of a shady cool creek on a hot day is a challenge of Herculean proportions.

Do not let the possibility of rounding up escaped pigs stop you from adding them to your homestead. Remember to feed them milk frequently and always from the same bucket, so they get to know it. If you are dealing with a stubborn pig, tie a rope around one back leg and pull in the opposite direction from the one you want the pig to go; you can now walk it anywhere.

The words *pig* and *hog* refer to the same animal, regional preference being the only difference. Swine refers collectively to the family Suidae, which includes pigs. Baby pigs are piglets and a piglet that has been weaned is a weanling; a young pig is a shoat, and feeder pigs are pigs from weaning age to market weight. A gilt is a young sexually mature female that has not had a litter; female pigs that have had a litter are called sows. Mature breeding males are boars; barrows are males which have been castrated at a young age, stags are males which have been castrated after becoming mature. To farrow means to give birth, and brimming refers to a sow or gilt in heat.

BREEDS

The pig has come a long way since it was first domesticated in China earlier than 4000 B.C. Even the pigs Columbus brought with him on one of his voyages are said to have weighed only an average of 170 pounds. Pigs are a clear example of improvement through selective breeding, as some mature boars today weigh over 1,000 pounds.

The various breeds can be separated into three general categories: lard type, bred to produce large amounts of fat; meat type, bred to produce a maximum amount of lean meat; and bacon type, bred to produce a maximum amount of both lean meat and fat in the belly, the bacon area.

There are more than three hundred breeds of pigs; the most common homestead breeds are the Durocs, Hampshires, Poland Chinas, and Yorkshires. Crossbreeds of these are also prevalent.

Durocs are a large meat-type pig, the boars and sows weighing 900 and 850 pounds respectively. This breed is red and has drooping ears, is known for its prolificacy and fast weight-gaining ability.

Hampshires are a little smaller, with boars weighing approximately 800 pounds and sows 650 pounds. Hampshires are a good meat type. This breed is black with a white belt over the shoulder and front legs, and has erect ears. The Hampshires are good mothers and are a hardy breed.

Poland Chinas, a meat-type pig, is one of the largest breeds, the boars weighing 900 and the sows 800 pounds. Poland Chinas are black with white on their faces, feet and tail tip, and have drooping ears. The Poland Chinas have fast weight-gaining ability, but are not as prolific or as good mothers as some of the other breeds.

Yorkshires are what most nonfarmers think of when they think of pigs, for it probably is the most widely distributed breed in the world. Yorkshires, a bacon-type pig, are white (actually their hair is white and their skin pink, so they look rather pinkish) and have erect ears. The Yorks are excellent mothers. If you live in a hot climate, Yorks are not a good choice unless you have plenty of shade; they are very susceptible to sunburn.

SELECTING PIGS

You should have at least two pigs, because pigs will eat better with competition at feeding time. Decide whether you are going to raise feeder pigs or if you are going to go into breeding stock. Whether breeding pigs is an economically sound route depends upon the equipment you have, the time available, and the amount of feed you can raise yourself. For the first year at least, you are better off raising feeder pigs.

Find a reputable breeder with the breed or crossbreed you want and look over the various sizes available. The smaller the pig the cheaper the cost, but the more feed you will have to provide. The time of year is a factor; if you live in a cold climate, the winter months are harder on the animal and you. Pigs consume more feed in the cold months and you will have to water them several times a day, as their water will freeze. Summer months are less work for you, but if you are raising your own feed corn it will not be ready until late in the season. It takes about six months for a pig to go from birth to a slaughter weight of approximately 230 pounds.

The cost of feeder pigs can fluctuate dramatically, as can the price of market-weight pigs. In our part of the country feeder pigs tend to consistently go for about $1 a pound, but the price of market-weight pigs has varied as much as 50¢ a pound in three months.

To pick out the best shoats, look over the sow and, if possible, the boar. A great deal of pork quality is hereditary. The parents should be alert and healthy-looking. Good pigs have a well-arched back, the high point at about the middle. They should have full, smooth shoulders, be wide at the hips, and have a high-set tail. The legs should be straight and not too thin, and the underside should be straight and firm, not overly plump.

Weight is an indicator of pig's quality; your shoats should weigh about twenty-five pounds at six weeks and from thirty-five to forty pounds by eight weeks. Pick pigs that are approximately the same size or the small one will not get a fair share of food and comfort. The gender of the pigs does not make much difference for feeder pigs, but if you get a male, he should be castrated; an uncastrated pig produces meat with an undesirable odor and flavor.

Weanlings from a reputable farmer should have had their needle teeth clipped, been wormed, and possibly had iron shots. Well-managed farms decrease your chance of buying pigs with hog cholera, viral pneumonia, leptospirosis, erysipelas, and other diseases.

SHELTER AND EQUIPMENT

If you are starting out with a couple of small pigs in the spring, your equipment needs are small and relatively inexpensive: a simple house, shade, a good fence, a feed trough, and a waterer. If you plan on keeping pigs year after year, you will need to rotate your pasture area each year (more frequently if it is small).

Summer shelters for pigs are the easiest to construct and the least expensive. A simple A-frame construction with a watertight roof and a back wall is most practical. Face the door away from the wind and put skids on the bottom so the house is movable.

The summer shelter can be even simpler if your pigs go to slaughter before the temperature drops below 50°F (10°C). You may then build just a shade shelter, which can be four-foot-high posts stuck in the ground and topped with a flat roof. Some loosely piled straw on the roof helps to control the heat. One warning, however: pigs love

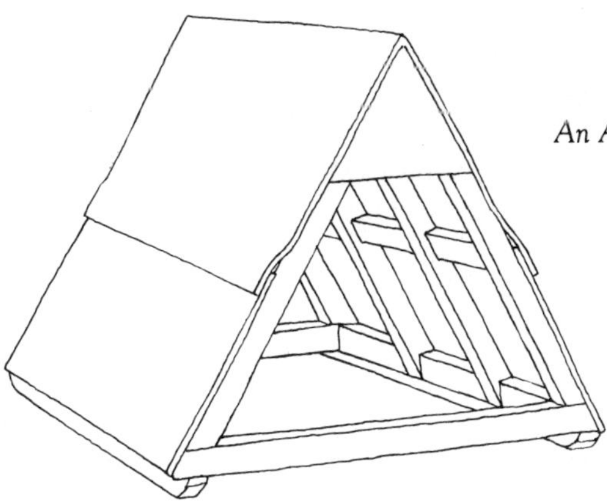

An A-frame shelter

to scratch themselves on anything handy, so whatever you construct for them must be very sturdy or you will soon be rebuilding it.

You may fence in a tree or two with your pigs for shade, but if the area is small they will probably uproot the tree or at best make a mess of the area around the trees. A couple of rows of sunflowers planted on the south side of the pen will cast a lot of shade by the dog days of August; do not plant them too close to the pen or the pigs will eat them.

Winter shelters for pigs in the northern part of the country need to be much more sturdy. The basic A-frame with three sides closed off will work. Keep the ceiling relatively low so that the pig's body heat can help heat the shelter. You do not need flooring; the ground covered with good clean bedding is much warmer.

Permanent shelters are excellent, but economical only if you already have an existing building to use. Completely confining the animals can be very successful, but it requires more work for the homesteader. A heated permanent shelter eliminates much of the risk if you are going to have farrowing sows in the cold months.

You may insulate your closed shelters, but the insulation must be covered by a vapor barrier. Most commercial insulation has a foil backing which is sufficient; if not, plastic sheets will work. The vapor barrier must be covered by a substantial wood siding on the inside; one-inch-thick hardwood is a good choice. Do not use plywood because the pigs will eat it! Ventilation is necessary in a closed shelter, and be careful to avoid drafts.

The portable pen is the type of housing we prefer for our feeder pigs. It is effective, economical, sanitary, and eliminates the need to fence in pasture. The pen is built of one-inch-thick boards and four four-inch corner posts outside them; the sides are four feet high. We placed a roof over a third of the top to provide shade and protection from rain. We have also built in a covered feeding box large enough to accommodate 100 pounds of feed to allow a free-feeding system.

This pen can be moved easily by placing small logs under the skids and sliding it along. Its very best feature is that once a week we move it the length of the pen so that the pigs have clean ground; they have left a nice piece of plowed and fertilized ground practically ready for planting. The height of certain sections of this year's corn is a perfect record of where the pigpen traveled last year.

The size of the area designated for your pigs depends upon the type and number of pigs you are keeping. Each feeder pig will need about 25 square feet of pen space on pasture. For a breeding sow or confined pigs, you will need about 100 square feet to allow sufficient room for piglets and exercise.

Unless you use a portable pen or tethering, you will need some type of fencing. As you construct your pig fence, keep in mind that pigs root with the hard ridge of cartilage along their snout, appropriately called the rooter.

Woven wire can be used for pig fencing; eleven-gauge fencing is excellent, but nine-gauge is good and more economical.

To make sure your pigs do not root their way under the fence, place logs against it on the inside of the pen or use a strand of barbed wire at the bottom. If you use barbed wire, have the bottom strand of the woven wire six to eight inches from the ground and put the strand of barbed wire approximately four inches from the woven wire.

You may also ring your pigs to prevent rooting. Rings and a ringer can be purchased from a farm-supply store for a modest price. Put one or two rings above each nostril on the edge of the rooter. Of course, ringing causes your pigs to be ineffective plowing machines for next year's garden.

Pigpens should be considered temporary constructions, as you will need to move them at least each season and optimally every six weeks to prevent the ground from becoming overly infested with parasites, keep the odor down and control the wallows (mud puddles).

Tethering eliminates the need for fencing and easily facilitates rotating the pasture area. To successfully tether a pig, you need a shoulder harness. You may use a dog harness or a cow halter, or make your own out of leather or cloth webbing. The tether should be a heavy chain.

A wallow can be anything from a constructed belly-deep wading pool to a nice mud area in the pen. Pigs do not wallow because of any desire to get dirty, but to keep cool. If the temperature goes above 90°F (32°C), give an area of the pen a good soaking with several buckets of water and you will have made friends for life.

We learned about pig wallows the hard way. The few times we have had problems with our pigs getting out of their pens have been because I did not supply them with a wallow on hot days. One time the little porkers got out four times in one day and each time headed for the horses' bathtub waterer. Horses do not like pigs too well, and I suspect ours are even afraid of them. It was pretty funny seeing the two pigs contentedly smiling in the bathtub and the two horses wild-eyed in the opposite corner of their paddock.

You need sturdy feed troughs for your pigs or they will destroy them in their exuberance at mealtimes. You can make a sturdy feed trough from scrap lumber; a V-shape aids in keeping the trough clean. A cut and filed steel drum makes an excellent feeder, too. If you feed slop, clean the trough every day so the remainder does not sour. Allow enough room for all of your pigs to eat at once. Plan on having at least $1\frac{1}{2}$ feet of space per pig.

Pigs need to have a lot of clean water. You may use a big bucket or a cut-down barrel, or construct a wooden watering trough. Make sure that the pigs cannot tip it over, especially in hot weather.

BREEDING

Breeding and farrowing are a bit more complicated with pigs than with other homestead animals. It definitely is a project that should not be undertaken until you have had the experience of a few seasons of raising feeder pigs. Pigs can farrow two or three times a year, but once a year when the weather is good is much easier.

You have a choice of keeping your own boar, taking your sow to somebody else's boar for servicing, or artificial insemination. Boars reach sexual maturity at around seven months, but should not be used too often until they are at least a year old. A boar maintains virility until he is about five years old. Some people keep a boar from an early litter, use him for breeding before he is a year old and then castrate him; a couple of months after castration, he is slaughtered. The advantage of using a young boar is that you do not have to house and feed a full-grown one. Older boars are not only huge, but may also be ill-tempered. In fact, one should always be wary around a boar in rutting season. You know when they are in rutting season because they start ranting, which means they run back and forth, snapping their jaws and slobbering.

Using somebody else's boar eliminates all this hassle. When your pig is in heat, take her to the boar you have selected rather than bringing him to her. (You can bring him to her, but first make sure you are equipped to handle an unfamiliar 900-pound ranting boar.)

Artificial insemination has the advantage of breeding your sow to the very best of boars without transportation problems. The minus side is that it is difficult to be sure when a sow is in heat. This factor makes artificial insemination rather impractical.

A gilt is ready to be bred when she reaches 250 pounds, at about seven to eight months. Choose a gilt that has twelve to fourteen teats and is the offspring of a good farrowing sow; heredity plays an important part in farrowing ability. A sow's farrowing ability will have already been tested, so you know what to expect.

The heat cycle for pigs is from eighteen to twenty-four days; a sow stays in heat for two to three days, while a gilt's heat period tends to be shorter. You can detect the heat period by observing her trying to mount other pigs and urinating more frequently. She may show some discharge from the vulva, which gradually turns pink and may begin to swell. During the early stages of heat the sow may become restless. Remember that detecting a brimming sow is not all that easy, and detecting a brimming gilt is even more difficult.

When you think you know your sow's heat cycle, record it on a calendar. Plan to take her to the boar when she is brimming again. Two weeks before she comes into heat, flush her (increase her feed just prior to breeding); this helps ensure the conception of a maximum number of piglets.

Take her to the boar at the first signs of brimming. Let her settle down before you have the boar service her. Conception cannot take place until the sow is in what is called standing heat; her vulva is swollen to its greatest extent and she has a definite discharge. She will stand solidly in position ready to be bred, hence the term.

Boars are so large and clumsy that you may need to assist with the mating to ensure conception. Large commercial operations use a breeding crate, but a couple of bales of hay on either side of the sow will accomplish the same objective: to support both sow and boar during mating. Bring the sow back to the boar, if possible, for a second breeding sixteen to twenty-eight hours later. After the mating takes place, be sure and watch your sow for any signs of coming into heat again about eighteen to twenty-four days afterwards. If there are no signs, she is probably bred.

The gestation period is approximately 114 days, which works out to 3 months, 3 weeks, and 3 days. During gestation the mother-to-be needs the best food. Feed five pounds a day of a 14 percent protein ration for the first $2\frac{1}{2}$ months and then increase to a 16 percent protein feed. After farrowing, the sow should have her feed gradually increased so that by the third or fourth day she is getting ten pounds of 16 percent protein ration per day, five pounds at each feeding. Keep this feeding schedule until you begin to wean the piglets, then gradually cut her back to five pounds of feed per day. Give her all the clean fresh water she wants, which will be from four to six gallons a day.

If you use a farrowing pen, put the sow in it three to five days before she farrows. The pen should be scrubbed down with a good disinfectant, such as a solution of $\frac{1}{2}$ pound of lye to 10 gallons of water. Before you put her in the pen, wash her down with a mild detergent and warm water. Make sure that her udder has been carefully washed.

A farrowing house is a dry, draft-free structure wherein a constant temperature of 70° to 75°F (20° to 24°C) is maintained at all times. As this is a little out of reach for most homesteaders, you will be reassured to know that the temperature range that piglets can withstand is from 40° to 90°F (4° to 32°C). In choosing when to breed

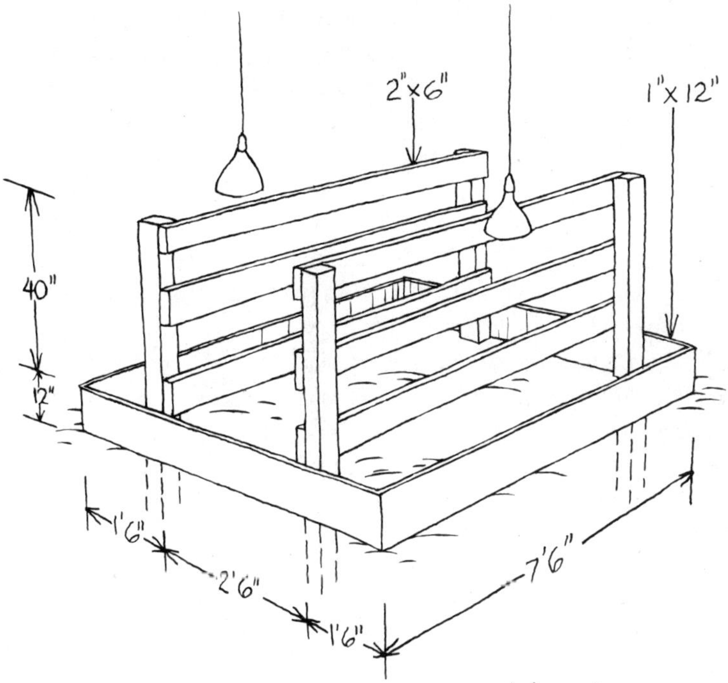

A farrowing pen

your pig, keep in mind the typical weather at the time she will farrow.

If you plan to have your sow farrow indoors, you may make use of a farrowing crate, a sturdy pen just large enough for the sow to lie in—about 3 feet wide and $7\frac{1}{2}$ feet long. The sides should be railing, not solid wood, so you can help out the sow and piglets if necessary. The bottom rail on either side of the crate should be 10 to 12 inches from the floor to allow the piglets to get out of mother's way. Sows are generally good mothers, but they are heavy, and with a large litter it is easy for some piglets to get stepped on or laid upon.

Provide the farrowing pen with a good straw bedding about two inches deep. You do not want it much deeper or it will be hard for the piglets to stay out of mom's way. Keep the pen clean; this may require changing two or three times a day, and can be done when you let the sow out for exercise. She should get out at least twice a day for fifteen minutes or so. Weather permitting, the sow and piglets can be put out on pasture when the piglets are seven to ten days old.

The simplest farrowing system for homesteaders is pasture farrowing. All you need is a three-sided shed with a good roof and floor to keep out rain or snow. Put straw in and near the shed so that the sow can build her nest when she is ready. Farrowing in the pasture eliminates the need for a farrowing crate, but choose your breeding time to ensure mild weather when the piglets arrive.

As farrowing time approaches, get your birthing kit ready so you can assist if necessary. You need clean towels, tincture of iodine, a small cup or jar, a pair of side cutters or nippers, petroleum jelly, and a pair of surgical gloves.

When the time arrives, the sow will become very restless and will busily build her nest. A sow will usually labor for fifteen to thirty minutes before expelling the first piglet; the other piglets will generally come faster. The afterbirth may come all at once after delivery or in pieces. Bury the afterbirth when it has all been expelled.

If the sow has worked for more than thirty minutes on delivering any piglet you should probably examine her. Before you do so, put on a lubricated surgical glove and enter the vagina slowly with your hand palm down and thumb tucked under the palm. If a piglet is coming head first, hold the back of its neck; if feet first, get the legs between your fingers and gently pull with the mother's contractions. That should get everybody moving out, but if not, call your veterinarian to assist you.

The sow probably will not be very hungry for about twenty-four hours after delivery, but she will need water. Give her all she wants. The first water you offer her may be warm, to help replace the lost body heat due to delivering. If you have a farrowing pen, try to feed the sow when the piglets are safely out of the way; they could get injured in her rush for food.

After farrowing, observe your sow for any signs of mastitis, metritis, or agalactia (see Medical Concerns, pages 156–159.

Optimal air temperature for newborn pigs is 85°F (30°C) for the first couple of days. If you are using a farrowing pen, use heat lamps with safety shields; the lamps are especially beneficial over the piglet area. Do not overheat the area where the sow is or she will get restless and be apt to injure the piglets. Your biggest job at farrowing time is to see that the baby pigs are out of the sow's way and that they do not get chilled. If they do, rub them briskly with a clean dry towel and put them under a heat lamp until the sow has settled down. If a piglet seems lifeless, try rubbing or slapping its sides to see if you can get it breathing. You should also dip the piglet's navel cords in tincture of iodine soon after they are born. Use your side cutters or nippers to clip off the tips of the eight tusklike incisor teeth, called needle teeth or milk teeth. If these teeth are not cut they are apt to damage the sow's udder. Sometimes they will not be present at birth, so check for them each day and cut them when they do erupt.

Piglets on their all-milk diet are very susceptible to anemia. Anemic pigs will not gain well, so protect your piglets by giving iron orally, by daily swabbing the sow's udder with an iron and molasses or honey mixture, or by giving an iron injection.

When the piglets are seven to ten days old you may start giving them grain. Use a creep feeder—a pen with small enough openings to keep the sow out, but large enough for the piglets—so the sow does not eat all their food. There are good pig starter pellets made by commercial feed companies. You might mix milk with the pellets at first to get the piglets interested (goat milk is excellent).

The piglets can be weaned at six to eight weeks, depending on their growth and how well they are eating grain. Do not wean them if they weigh less than twenty-five pounds. Reduce the sow's rations two or three days before weaning time to reduce her milk flow. If you are going to raise only one litter a year, let the piglets stay with her until she dries up herself; this eliminates the need to separate the sow and piglets, which can be a problem on a small homestead.

FEEDING

Pigs are monogastric (single-stomached), so their feeding needs are different from ruminants, such as sheep and goats. Commercial feed for pigs is available at all stages of their growth and development. This grain is the easiest to feed, as it is fully prepared and offers a balanced diet. A disadvantage of commercial feed is its cost. A good compromise is to use commercial feed for the basic diet and your own products as supplemental feed.

Grains grown on your farm can be used as feed. Corn is the chief feed for pigs, but by itself it is not an adequate diet because it is only 8.6 percent protein. Wheat, oats, and barley are higher in protein than corn, but more expensive. Hard wheat is 15.8 percent protein, soft wheat 9.9 percent, and oats can be up to 12.7 percent. Ground oats and barley can be fed as one-fourth to one-third of the grain ration.

Milk and whey are excellent products to feed your pigs. A feeder pig can consume 1 to $1\frac{1}{2}$ gallons of milk per day. Not only do pigs thrive with the addition of milk to their diet, but milk aids in controlling internal parasites.

Garden surplus particularly palatable to pigs includes comfrey, Jerusalem artichokes, lettuce, cabbage, potatoes, and just about anything else you grow. In fact, your pigs and garden can get together for a gleaning or hogging down. Gleaning means putting your pigs in a field after it has been harvested so they can clean up anything that has been left. If you lose a crop before it is harvested due to the weather, you can have your pigs hog it down by turning them out on it as soon as the ground dries up sufficiently. If you use your pigs to hog down a field, make sure you get them adjusted to the change in diet slowly so they do not get colic.

Swill is any available foodstuff that is thrown into a kettle, cooked, and then fed to pigs. Swill is cooked to destroy any parasites in the foodstuff. Clean garbage from your kitchen, such as potato peels and meat scraps, can be prepared in this manner. Make sure the swill has cooled down before you feed it.

Some people have successfully fed garbage to their hogs. You can make arrangements with a restaurant to pick up its garbage on a daily basis and feed it to your pigs. However, you run a high risk of losing your pigs this way, since you cannot be sure that they are getting a proper diet. You also are depending upon the garbage's not containing any cellophane, knives, or broken china. Remember too that any infected undercooked pork that is eaten by your pigs can transmit trichinosis to your animals.

Pasturing pigs is an excellent system for feeder pigs if you have the available pasture. One acre of good swine pasture can support eight to ten pigs. You can get special mixtures of pig pasture crops that contain just what your pigs need. The best pig pasture crops are alfalfa, rape, red clover, and Ladino clover.

If you are not feeding commercial pig feed, protein supplements should be fed to pigs until they weigh at least 100 pounds. The supplement can be soybean-oil meal, meat scraps, milk or milk by-products, or tankage, which is made by rendering meat or fish by-products into a feed containing 55 to 60 percent protein. The amount of protein supplement needed will be less than $\frac{1}{2}$ pound per day per pig for tankage or soybean oil, or 1 to $1\frac{1}{2}$ gallons of milk or milk by-products. Before you elect to use a feeding plan requiring a protein supplement, determine the ease of obtaining the supplement on a regular basis.

Salt should be offered to your pigs unless you are feeding commercial pig feed. Supply a salt block in a protected box in their pen or mix about $\frac{1}{2}$ pound of salt with each 100 pounds of feed you prepare.

As your pigs approach slaughter weight, you are said to be finishing them. Pigs weighing more than 225 pounds are eating more than they will return in meat cuts, so many people feed just corn or corn mixed with commercial pig feed the last four to six weeks to finish the pigs off. This adds fat to the pig, and the flavor of meat comes from the fat.

While finishing your pigs, avoid feeding much soybean or peanuts, as they cause soft pork. Soft pork is edible but has a very mushy texture. Also avoid feeding highly flavored foodstuffs, such as onions or garlic, that might leave a flavor in the meat.

One problem you will have until you develop an experienced eye is judging your pig's weight. You may buy a swine "weigh-tape" which helps determine a pig's weight by measuring its length and the heart girth (chest circumference). Instructions come with the tape.

COSTS

Your pig will consume between 600 and 700 pounds of feed from weaning to slaughter weight. If you are feeding only commercial pig feed, you can determine your costs ahead of time by adding the purchase price of your pigs to the cost of 650 pounds of pig feed. If you plan to have your pigs slaughtered commercially, add that cost. You will get about 150 pounds of meat cuts from a 225-pound pig, so you can estimate your total cost per pound of edible meat.

CASH COSTS OF RAISING A PIG

Feeder Pig (40 pounds at $1 per pound)	$ 40.00
Cost of Feed (650 pounds at $9.60 per 100 pounds)	62.40
Slaughter costs ($15); cutting and freezer wrapping (10¢ per pound)	30.00
Total Cost:	$132.40
Cost per pound of edible meat:	$ .92

Of course if you can reduce any of the above costs, the cost per pound for your home-raised pork will be lower.

MAINTENANCE

The general rule for maintaining healthy pigs is to keep them in clean, dry pens and feed a good balanced diet. You should make frequent checks for signs of lice and mites; if evidence of these parasites is found, treat the pigs with appropriate powders and sprays purchased from any farm-supply store. Lice are small, grayish in color and can be seen by the naked eye; an infected pig will scratch almost continuously and act restless. Mites cannot be seen; they burrow into the skin and cause mange. If your pigs constantly rub their backs and sides, suspect mites. If mites are left untreated, the pigs will rub their hair off and red, scabby areas are apt to appear.

Worms affect pigs. Good sanitation and pasture rotation does a lot to control the situation. There are broad-spectrum worming medications that you may purchase and use according to directions. If your animals are not doing well, have a veterinarian do a fecal examination to make sure they do not have a type of worm unaffected by your wormer. Always worm your sow before breeding her.

Vaccinations are available for pigs. Advisability of administering various vaccines can be determined by checking with your veterinarian or agricultural extension service to find out the incidence of various diseases in your area. There are vaccines available for erysipelas, hog cholera, transmissible gastroenteritis, and leptospirosis.

Castrating any boar you plan to use for meat is essential. Baby pigs may be castrated from one day old up to four or five weeks; it is best to do it when they are about a week old. At that age they have gotten a good start, but are still easily handled.

You need a pail of warm soapy water; tincture of iodine; a clean towel; a single-edged razor, scalpel or a sharp knife; and somebody to help you. Have your helper hold the pig off the ground by the hind legs with its belly toward you (the person holding the pig can effectively use his or her knees as a brace). Scrub the area in front of the scrotum with soapy water and dry it. Push the testicles toward the belly and forward as far as they will go. Make an incision about an inch long between the testicles. Squeeze one testicle out as far as the cord will allow. If the testicle won't come out, lengthen your incision and try again. If the pig is young enough, the cord will snap on its own; if it is an older pig, you will have to scrape the cord. Do this by cutting the cord with a scraping motion to minimize bleeding. Cut the cord close to the body. Remove all of the white membrane around the testicle and make sure none is left sticking out through the incision. When you are finished, pour tincture of iodine on the incision.

Occasionally one finds a ruptured pig. In a normal pig the testicles are definite and hard in the scrotum; a ruptured pig has a large scrotum filled with a soft material as well as the testicles, because its intestines have descended into the scrotum. A pig may be slightly ruptured, ruptured on one side or both, or severely ruptured. Sometimes pigs rupture after you castrate them, so check for this.

If you have a ruptured pig you may leave him to sell for slaughter after weaning, call your veterinarian to repair the rupture, or plan to sew him up yourself. If you are going to do it, use a darning or carpet needle and heavy thread. Be careful not to put the needle into any intestines while closing the two sides of the scrotum incision. Use two or three stitches, making sure you have not left any opening for the bowels to work through. Fasten the ends securely but not too tightly. Finish the job by dowsing the area with tincture of iodine. Keep a check on the incision to make sure it has not become infected. Remove the stitches when healing is complete, after about fourteen days.

Castrate larger boars (over 100 pounds) the same way as baby pigs, but tie off each testicle to prevent bleeding. Tie the cord with catgut and pull it as tight as you can. Wrap and knot the catgut several times, then cut the cord on the testicle side of the tie. A good strong rope and assistants are not luxury items in castrating a large boar! Remember to feed a boar at least a month after castration before you slaughter him for consumption.

MEDICAL CONCERNS

There are a number of pig diseases, which is why you should start with healthy stock, maintain a clean pigsty, and a good diet. A pig's normal temperature is 102°F (40°C).

Anemia in baby pigs has already been mentioned, but since an average of one out of every four piglets dies before weaning age, it is worth mentioning again. Give iron injections or oral iron to help prevent anemia.

Atrophic rhinitis often starts with scrapes or bites. The symptoms are persistent sneezing and nasal discharge. This disease can spread, so call your veterinarian for assistance in diagnosing the specific organism causing the problem. Keeping the pig's snout clean is a good preventive measure.

Brucellosis can affect pigs, though it is caused by a different organism than the one which affects sheep, cattle, and goats. The first sign of brucellosis is usually infertility, abortions, or stillbirths. You may also notice lameness or paralysis. There is no cure. Pigs can be tested for brucellosis, and you should buy pigs from tested, brucellosis-free herds. Brucellosis can affect people as well; it is contracted by eating undercooked pork or by butchering with an open wound on the hands.

Hog cholera (swine fever) is caused by a virus and is almost always fatal. The symptoms are a sudden onset of fever, loss of appetite, weakness, and heavy consumption of water. If you suspect hog cholera contact your veterinarian at once.

Mastitis is an inflammation of the milk-producing glands caused by bacteria. The area around the teats becomes very hard and hot, and is painful when touched. The sow may run a temperature of up to 107°F (42°C) with this condition. Call your veterinarian. Mastitis is treated with a broad-spectrum antibiotic; apply hot packs and use a warmed udder ointment. You may also try to "milk" that section of the udder.

Metritis is an inflammation of the uterus; if it occurs, it generally appears within thirty-six hours after farrowing. The symptoms are a white to yellowish discharge from the vulva, a temperature of up to 106°F (41°C) and a continuing depressed appetite. Call your veterinarian, who will treat this condition with a broad-spectrum antibiotic.

Agalactia is a reduction in or a lack of milk flow; it usually occurs with mastitis and metritis. The sow will have a depressed appetite, may run a temperature of 106°F (41°C) and the udder will feel hard and feverish. Again, you need veterinarian assistance if this condition occurs.

Erysipelas is caused by a bacteria and can also be transmitted to people. The symptoms are similar to hog cholera, including elevated temperature, purplish blotches over the body, depressed appetite and swelling of the nose, ears and limbs. If diagnosed early enough, treatment is often successful. Call your veterinarian.

Heatstroke is caused by hot weather, too much exercise, lack of shade, and lack of water. The animal suffering from heatstroke will pant excessively and perhaps stagger and fall over. To treat an affected pig, put cool water on its head, legs, and body. Do not chase your pigs if they get loose on a hot day (chasing escaped pigs is not a very effective system for capturing pigs anyway).

Leptospirosis is caused by a bacteria. It can affect people, dogs, cattle, wild animals, and rodents as well as pigs. Abortions and stillbirths are symptomatic of this disease. Affected hogs have yellowed gums and eyelid membranes. If you suspect leptospirosis, call your veterinarian. The organism is passed in urine, and since rodents can transmit it, be sure to keep your feeds covered.

Pneumonia can be caused by a virus or bacteria; an affected pig has a depressed appetite, breathes hard, runs a high temperature and acts lifeless. Isolate the affected pig in a dry draft-free pen and call your veterinarian.

Scours is diarrhea in baby pigs; their stools become loose and yellow. To treat, decrease the sow's feed to decrease her milk production. Clean the bedding to prevent the pigs from catching cold. You may provide extra water and you may also give the piglets a small dose of a commercial antidiarrheal agent.

Feeder pigs on the homestead can be a distinct asset. You can produce 150 pounds of meat for less than $1 a pound in under five months, with a minimum of effort on your part. Your reward will be pork that is delicious and hams with no water added—try to find that kind of meat in the supermarket!

Pigs lend themselves well to a sharing system among homesteaders. You can raise two pigs, one for yourself and one for a homesteading friend, and the next time around have the other family do the raising. You also may raise two pigs and sell one for a profit if you do not have room in the freezer for two pigs at once. There are quite a few possibilities, and besides, a homestead just does not seem like a homestead without a pig or two around.

horn
poll
tail head
pin-bone
rear udder attachment
rear udder
dewlap
brisket
milkwells
flank
fore udder
stifle
thigh
mammary veins
teat
hock
switch
pastern
hoof
sole
dewclaw
heel

7

CATTLE

MOST homesteaders sooner or later give some thought to raising a beef animal or trading in the goats for a family cow. The addition of a member of the bovine family to your homestead requires careful consideration. However, even if you never plan to own a cow, we suggest that you read this chapter if you own goats. Every goatherd will sooner or later run into a cow person, and you will fare better in conversation if you can speak the language.

The efficiency of the milch cow in converting feed into food energy for people is excellent, but a beef animal's conversion level is the lowest of all the meat animals. The feed:product conversion ratios are 2.8 pounds of feed per gallon of milk and 20 pounds of feed per pound of beef.

The most obvious fact about cattle is that they are big animals and therefore require a great deal of space and feed. Raising or maintaining a cow should not be considered unless you have adequate pasture, both in quantity and quality.

When we first moved into our homestead we watched our neighbor, Len Wright, lead his black Jersey out on nice days to be tethered under the old apple tree. She had not freshened (given birth) in more than eight years, but she was still giving more than a quart of milk a day. The milk she gave provided all the milk and butter the family needed, even if it was not economical. There was a lot of country romance surrounding Wright's cow, and when she was finally sold the neighborhood just did not seem the same.

Our Katie had been quite fond of Wright's cow, and after it was sold she started talking about wanting a little cow for her birthday. Her request did not go unnoticed, and in time for her fifth birthday, Feline, a month-old Jersey heifer, arrived. She was a beautiful animal with golden tan coloring, a little white around her nose and gorgeous brown eyes. She was a lot of fun and very affectionate. She would follow us around like a puppy and would constantly lick us if we let her. As Feline approached her full size and weight, Katie started wishing she did not have a pet cow. A playful nudge from Feline did not seem too pleasant to a forty-pound little girl. Meanwhile, Dad had been reconsidering the wisdom of owning a cow as he hauled the bales of hay, and Mom was getting panicky about handling forty to fifty pounds of milk a day after the cow freshened. We sold her to our friend Nap, who loves her dearly—and has lots of pasture.

Cattle refers to both genders of various animals of the genus *Bos*; they are in the Bovidae family, as are sheep and goats. Cow refers to a mature female that has freshened at least once; a young animal is called a calf, and a newly born calf is called a deacon. A heifer is a young female and a bull calf is a young male. A steer is a castrated bull being raised for meat, and a stag is a bull that is castrated after reaching maturity. A castrated bull used for draft work is called an ox. Young beef cattle not at their full weight are called feeders or stockers. Beeves is the plural of beef, and you may therefore hear a herd of young beef calves referred to as beeves.

A vealer is a calf up to 250 pounds that is sold for veal; bob veal is a marketed newborn calf. Veal calves are generally dairy calves. Baby beef is meat from beef calves marketed at ten to fifteen months of age at a weight of 650 to 850 pounds.

Bull means an uncastrated male, and there are beef bulls and dairy bulls. Beef bulls, despite their enormous size and weight—2,000 pounds or more—are generally docile animals (with the exception of the Brahman and Brahman cross), but dairy bulls are among the most cantankerous and dangerous animals alive. The disposition of the dairy bull is the primary reason for the development of artificial insemination.

BREEDS OF BEEF CATTLE

Breeds of beef cattle include Angus, Red Angus, Devon, Galloway, Hereford, Shorthorn, Charolais, Highland, Brahman and Brahman crosses. The most appropriate breeds for the homesteader are Angus, Hereford, and Shorthorn.

The Angus, a black polled (hornless) breed, originated in Aberdeen, Scotland, and is considered hardy. The Angus matures early and produces a high-quality carcass with good marbling. They have a compact, broad, and low-set body.

Herefords can be either polled or horned and range from medium to dark red with white face, flank, underline, breast, crest, and areas below the hocks and knees on both the front and back legs. The Herefords originated in Hereford, England, and are known for their hardiness and foraging ability. The body conformation of the Hereford is muscular, with well-developed back, loin, and round.

Shorthorns can be either polled or horned and are red, roan, white, or red and white. The breed originated in the northeastern section of England and were originally considered dual purpose—for both milk and meat production—but most have been bred for meat. Shorthorns are noted for a quiet disposition and rapid gains in feed lots. Their body conformation is compact and low-set.

SELECTING BEEF CATTLE

In selecting a breed of beef animal to raise, the homesteader should consider the family's personal likes, availability of the breed, environmental conditions under which the animal will be raised, and kind and amount of feed available. Specific considerations to keep in mind are the animal's age, gender, grade, general conformation, and health.

Steers or heifers are usually purchased in the fall of their first year or in the spring after they are a year old. The price depends upon the weight, so the older the animal, the more expensive.

It makes little difference whether you buy a heifer or a steer for the meat animal, although heifers generally do not require as long a fattening period as steers. If you buy a male, be sure he has been castrated.

Feeder cattle are graded according to their overall conformation. The grades are prime, choice, good, standard, commercial, utility, and inferior. The last three grades are not worthwhile for the homesteader to raise. Prime cattle make up a very small percentage of feeder steers and heifers. Prime grading goes to those animals showing exceptional smoothness and body conformation; they are very expensive.

Choice cattle are superior in conformation and quality. They have long bodies, moderate depth, long rumps, and well-developed hind quarters. Good-grade cattle show less muscling and lack development in the hind quarters, back, and loin. Good feeder cattle may reach choice grade after finishing (fattening). Standard grading indicates uneven top and underline, prominent hip bones, and a tendency to be narrow over the back and light in the hind quarters. Standard-grade steers or heifers should not be expected to finish off higher than good.

General good health is a must in the animal you select. Choose a calf that acts alert, is bright-eyed, active and shows no sign of scours (diarrhea), evidenced by a pasty tail or dirty hind quarters. Check your prospective animal against signs of swollen knees or poor hoofs. Avoid calves that demonstrate signs of rapid or difficult breathing, which could be a sign of pneumonia.

SHELTER AND EQUIPMENT

The best shelter for beef cattle is a three-sided structure with the open side facing away from the prevailing winds. Cattle grow a good heavy coat of winter hair that protects them through the coldest of winters if they can avoid cold winds. You actually do not need any shelter if you have a good natural windbreak and the weather is not too severe, but feeder cattle gain faster and more efficiently when protected from extreme weather conditions. Shade, either natural or artificial, is a must during hot weather.

The floor of an outside shelter should be dirt for insulation, with a layer of straw over it. If you house beef cattle inside, the barn should not be kept too warm. If the barn is much over 50°F (10°C) in the winter, the humidity rises because of accumulated urine, manure, and body moisture, and pneumonia can result. The barn should be adequately ventilated. The floor-space requirements are twenty-five to thirty square feet for a weaned calf and thirty-five to forty square feet for a yearling.

Cattle housed in a barn will require more time, as the manure will need to be cleaned up and the bedding changed frequently. You may use a deep-litter system (letting the bedding and manure build up until spring), but this procedure is backbreaking when it is time to dung out, unless you can use a tractor.

A hay and grain rack designed to permit self-feeding will save you a good deal of time. Any feed dispenser used outdoors should have a roof to protect the feed from the elements.

Cattle require a constant source of fresh water. A mature animal may consume up to twelve gallons of water a day, so devise a system that is sufficient for the animal and manageable for you. A continuously running stream through the pasture is ideal, but a water tank equipped with a self-contained electrical heating unit for cold weather is fine.

Good sturdy fencing for cattle may be made from wood, woven wire, barbed wire, electric fencing, or a combination. Woven wire should be strung with the posts about sixteen feet apart. Barbed wire should be strung taut, ten to fourteen feet between posts and four strands high. Neighbors and passing motorists tend to get uptight about finding cattle in their way, and rounding stray cattle up rates low on a scale of fun ways to spend an afternoon. Make your fences secure and frequently check them.

FEEDING

Years ago, beef cattle were usually raised to two or three years old before they were slaughtered. You may still do this, using cheap pasture and roughages without concentrates during the winter, if you have an ample supply of pasture and roughage. More economical gains are made, however, when cattle are fattened at a young age. Calves raised for early slaughter must be fed liberal amounts of grain and concentrates to ensure fattening along with their normal growth, which increases your capital investment.

To determine the weight of your animal for assessing weight gains, use a "weigh-tape" for cattle. The tape allows you to determine weight by measuring the animal's length and heart girth (chest width). Instructions for use come with the tape.

Calves are best left on the cow until four to six months, but they can be weaned from milk at thirty to sixty days if they are consuming hay and $1\frac{1}{2}$ to 2 pounds of concentrates (such as Calf Starter) per day. Calves will begin to nibble on high-quality grass hay and concentrates at about a week old. As they fatten, increase the milk and concentrate intake. A general rule of thumb for feeding milk or milk replacer is to feed one pound of milk for every ten pounds of body weight. Restrict the concentrate ration to three pounds a day to encourage consumption of roughage.

Calves being carried through the winter to be fattened later may be fed entirely or primarily on roughage, if it is of good quality. A pound of high-protein supplement per day may be used to ensure adequate gain; a gain of $\frac{3}{4}$ to 1 pound per day is desirable.

The rations fed to fatten cattle are determined by the animal's age and weight. Fat yearlings fifteen to eighteen months old weighing 650 to 850 pounds may be put on pasture after wintering until midsummer or fall, and then on grain with good roughage for sixty to ninety days. This method maximizes hay and pasture and minimizes grains.

If pasture and hay are abundantly available at a good price, you

SAMPLE RATIONS
FOR WINTERING A BEEF CALF

Ration Plan	*Type of Feed*	*Daily Amount*
1	High-quality hay, at least half legumes	12–15 lb.
2	Corn silage	25–35 lb.
	High-protein supplement	1 lb.
3	Good grass hay	10–12 lb.
	Crimped oats	2 lb.

may want to finish your animal at eighteen to twenty-four months weighing 850 to 1,100 pounds. Such animals are generally held through the first winter on hay alone, low-quality hay and one pound of protein supplement, or two to three pounds of grain. In the spring they are put on pasture alone, and in the fall they are finished on hay and grain for 90 to 120 days.

When you change to a finishing ration, do so very gradually to prevent digestive disturbances. If your animal is being fattened after coming off of total pasture feeding, start with a grain ration of 2 pounds per day fed in two feedings. Gradually increase the grain ration by no more than $\frac{1}{4}$ to $\frac{1}{2}$ pound per day, as the animal shows interest in additional grain. It may take four to six weeks to reach the full grain ration.

The best grain to feed your beef animal is the grain that is the most readily available at the best price. Corn is the most commonly used, either cracked or meal. If the hay fed with the corn is not a high-

FINISHING RATIONS

Age/Weight	*Type of Feed*	*Daily Amount*
15–18 mo.	Legume hay	6–8 lb.
650–850 lb.	Grain	14 lb.
18–24 mo.	Legume hay	10 lb.
850–1,100 lb.	Grain	15 lb.

quality legume, add a 10 percent ration of protein supplement. Oats are used particularly at the beginning of the graining period because of its bulk. As other grains are added to the ration, reduce the oat proportion. Oats should be crimped.

You may buy a complete pelleted dry feed for beef cattle. This product reduces the amount of time and effort expended, but the cost is high. Compare local prices for complete pelleted rations with grain prices to see if this feeding plan is worth your consideration.

Land that is unsuited for forest or garden is ideal for grazing beef cattle. The quality of a pasture can vary considerably, but it is generally divided into two classes, legume and grasses. The best pasture for cattle is a legume-grass combination, such as alfalfa and broom grass. An all-grass pasture is low in protein and an all-legume pasture is apt to cause bloat.

The amount of pasture you need to support a beef animal depends on the field's quality. You will need about $1\frac{1}{2}$ acres of good pasture per head to start; make adjustments according to your animal's gains.

Salt should be fed free choice in either block or loose form. Cattle should have access to salt at all times. Mineral deficiencies are common, particularly if a low-grade hay forms part of the ration. In this case, provide a mixture of trace mineralized salt, ground limestone, and dicalcium phosphate free choice in protected dispensers.

COSTS

The cost of raising a beef animal varies according to the feeding system you use and the size to which you raise your animal. The cost of beef on the hoof has fluctuated dramatically in the past several years and so it is impossible to give an accurate cost. The basic formula will be the same, however, and you can check on beef and grain prices in your locale.

Our calculations assume you own or have access to about $1\frac{1}{2}$ acres of fenced, good-to-high-quality pasture; low-cost, high-quality hay;

Cost of Raising Beef Steer

1 Steer, 450 lb. @ 44¢ per pound	$198.00
Feed Costs	
Winter rations (183 days): 12 lb. Legume-grass hay per day @ $80 per ton	86.04
Summer rations (123 days): pasture, average gain $1\frac{1}{2}$ lb. per day	—
Finishing rations (90 days):	
Pasture, hay, and grain, average gain $2\frac{1}{2}$ lb. per day	—
Hay, 8 lb. per day ration @ $80 per ton	28.80
Grain, starting at 2 lb. per day, increase of $\frac{1}{2}$ lb. per day every two days to full-feed ration of 14 lb. per day; 950 lb. @ $11.40 per 100 lb.	108.30
	$421.14
Slaughter cost	10.00
Cutting and freezer wrapping, 450 lb. @ $.15 per lb.	67.50
Total Cost:	$498.64
Value of Product	
1 beef steer, 900 lb. live weight, yielding 450 lb. of meat @ $1.11 per lb.	

The per-pound price compares very favorably to what you pay in the store for a pound of any cut of beef, but remember that the price per pound does not reflect a whole year's worth of your time and effort.

and sufficient storage space. Without these advantages, we do not recommend that you raise a beef animal. If you still hanker for some good beef not loaded with hormones, you might join forces with somebody who is equipped to raise beef animals or buy local beef from a nearby farm or slaughterhouse. We further assume the purchase of a good choice grade animal in the fall of the year, to be kept until approximately eighteen months of age, 900 pounds live weight, the following fall.

BREEDS OF DAIRY CATTLE

Adding a dairy cow to the homestead can be practical if you have a large family of big milk drinkers and if you enjoy making milk products such as butter and cheese. The cow will give you what a goat will, only in much larger volume. She also requires more feed, space, time, and effort than a goat. Whether a milch cow is for your homestead or not depends upon your preference and setup.

The five common breeds of dairy cattle are the Ayrshire, Brown Swiss, Guernsey, Holstein-Friesian, and Jersey. The Ayrshire, first developed in Scotland, is red with white markings or white with red markings. Ayrshires are inclined to be an active breed and may tend to be a bit nervous. The mature cow weighs about 1,200 pounds and gives an average of forty pounds (twenty quarts) of milk a day. The milk averages about 4 percent butterfat content.

The Brown Swiss breed was developed in Switzerland; it is the largest and most meaty breed of dairy cattle. Its color runs from light fawn to a very dark brown with a light-colored stripe along the backbone and muzzle. Brown Swiss are generally quiet and easily managed and are noted for their ruggedness. The mature cow weighs about 1,300 pounds and gives an average of forty-two pounds (twenty-one quarts) of milk a day, averaging about 4 percent butterfat content. Brown Swiss calves make excellent veal animals or can be raised for beef.

The Guernsey breed was developed on the island of Guernsey in the English Channel. Its colors range from light fawn to red, with white markings on the face, flank, legs, switch (tail tip), and sometimes on the body. Guernseys are an active breed, but are not nervous and are easily managed. The mature cow weighs about 1,000 pounds and gives about thirty-eight pounds (nineteen quarts) of milk a day, with a 4.7 percent butterfat content.

The Holstein-Friesian breed, usually called Holsteins, originated in Holland; they are the familiar black-and-white cattle. They are a large, rugged breed, with mature cows weighing 1,250 pounds; the cows of this breed are generally quiet and docile. You may expect a Holstein cow to give about forty-eight pounds (twenty-four quarts) of milk a day. They are the largest milk producers, but have the lowest average butterfat content, 3.6 percent. Holstein calves make excellent veal calves.

The Jersey breed was developed on the island of Jersey in the English Channel. The cattle range in color from light fawn to black and may have spotted or solid white markings. This breed is active and can be quite nervous. The mature Jersey cow weighs about 900 pounds and gives milk of superior quality: an average of thirty-four pounds (seventeen quarts) of milk a day with an average butterfat content of 5.3 percent. The Jersey is the most common homestead cow; its popularity is undoubtedly due to its size, lower feed demands, and superior milk quality.

There are three relatively common breeds of cattle known as dual-purpose breeds: Milking Shorthorn, Red Poll, and Dexter. The dual-purpose breeds do not give as much milk as the diary cattle, nor do they produce as much meat as a beef animal, but they do provide a good compromise. Calves from a dual-purpose animal need not be reared as vealers, which has potential for the homesteader interested in raising a beef animal.

SELECTING A DAIRY COW

The selection of a breed of cow for your homestead depends broadly on the amount of pasture available, outlets for milk produced, family preference, desirability of raising a veal or beef calf, and the availability of the preferred breed in your area. Selection of a particular cow is based on her general conformation, health, age, disposition, and production record.

A desirable cow has a smooth gait and good carriage, her body parts in proportion to each other. Her udder should be symmetrical, wide and deep, and strongly attached, with some cleavage between halves. The quarters should be evenly balanced. The teats should be a uniform size, placed squarely under each quarter with good spacing, and of medium length (neither pendulous nor small) and width.

The animal should look healthy: bright-eyed, alert, and active. Find out the animal's health record to ensure it has been free of mastitis, udder trouble, and lameness. Do not buy a dairy animal that has not been recently tested for Bang's disease (brucellosis), leptospirosis, and tuberculosis; it is best to buy a cow that has been vaccinated against both leptospirosis and Bang's disease.

The peak years of milk production for dairy cows are from seven to nine, and the family cow will continue to produce milk at an economic rate, allowing for individual differences, until about fourteen or fifteen. High-producing commercial cows (some giving up to 100 pounds of milk a day) are replaced at about six years old.

A cow that has had at least two freshenings can provide a better indication of her production ability. A first-time freshener is not a good buy unless you know her heritage of productivity and you are an experienced milker. A cow at first freshening must learn to be milked, and the situation is compounded if you need to learn how to milk at the same time. If the cow is milking at the time of purchase, be sure to find out the date of her last freshening, if and when she has been rebred, and be sure to sample her milk to see if you like the taste.

Your prospective cow's disposition is very important. A mean or nervous cow has no place on a homestead, particularly if children are going to be working with the animal. If possible, milk the cow before buying her to ensure a docile nature.

If you are buying a cow from a commercial herd, ask to see her production records. Look for the number of lactation days and pounds of milk produced during the cow's lactation period. Owners of a noncommercial family cow should also have records on the cow's production level. If the cow is milking at the time of purchase, be on hand to milk, and weigh and taste the milk. Milk-production level is measured in weight rather than volume, since the foam can be very deceptive. Remember that the amount of milk produced per day depends upon where she is in her lactation period.

CATTLE

SHELTER AND EQUIPMENT

Dairy cattle, like beef cattle, do not need elaborate shelter. You probably require more adequate shelter for a milch cow than for a beef animal, because you have to milk her twice a day. Electricity and running water in the shelter or barn is a tremendous asset.

Your barn will need a good ventilation system and should not be allowed to get much warmer than 50°F (10°C) in the winter. A warm barn, combined with the high humidity from the cow's breath and urine, produces an ideal situation for the development of pneumonia.

The stall area must be kept as clean as possible to cut down on the time you need to clean her off before milking. Cows do well in a stanchion stall $3\frac{1}{2}$ to $4\frac{1}{2}$ feet wide and $4\frac{1}{2}$ to $5\frac{1}{2}$ feet long, depending upon the size of the cow. Manger and waterer space will need to be provided at the front of the stall, and a gutter for manure and urine at the end. Gutters are typically 16 inches wide and 10 inches deep, but a single cow will do well if the stall is just a raised platform and the waste material drops to the main floor level. The waste should be cleaned up daily.

You also need sufficient space to keep her calf when it comes and adequate space for storing feed and hay. Allow three to four feet of extra space in the shelter as a walkway for your cow to get out of her stall and to the outside.

The stanchion can be a metal one purchased from a supply store, a wooden one constructed at home, or a wood or metal pole to which the cow can be secured by a chain attached to her collar or halter. Whatever type of stanchion you use, make sure it allows the cow to have a comfortable head position when she lies down. Some cow owners leave their animals in their stanchions through cold weather, in which case it is a good idea to see that they get some exercise outside when the weather is not too severe.

To make sure your cow is motivated to go outside for exercise even during the winter months, place a hayrack outside. A hayrack also greatly reduces the amount of wasted hay.

A cow must have fresh water available at all times, so provide a watering system both inside and outside the shelter. The combined waterer sizes should be adequate to supply a cow's daily need—up to fifteen gallons.

Dairy cows need salt, which can be provided loose or in block form. Provide salt in the pasture when in use.

Fencing needs for dairy cattle are the same as for beef cattle, as described on pages 164–165. Many homesteaders tether their dairy cows. Use a fifteen- to twenty-five-foot lead and change the cow's location frequently enough to provide adequate pasture. This method is highly successful, as it reduces the danger of parasite infestation of the pasture area and eliminates the need for yards of expensive fencing. The major disadvantage of tethering is that it requires some type of shelter or confined area in which to put the cow during inclement weather, at night, and when you are away; you, of course, have to lead the cow to and from the shelter.

FEEDING

A cow is usually fed twice a day, with the grain fed while she is being milked. The bulk of dairy cattle rations are made up of roughage such as pasture, hay, or silage. The type and amount of pasture requirements are the same for dairy cattle as for beef cattle. The type and amount of grain is determined by the type and amount of roughage, the amount of milk produced, percentage of butterfat, and the cow's weight. Due to these variables, there is no simple formula to use in determining adequate rations for your particular cow if you are going to mix your own grains; you would be wise to call in your county extension agent to help you determine the best and least expensive formula for your cow. The extension agent will also help you determine the quality of your pasture and help you to maintain or improve it.

A less complicated but more expensive feeding system uses commercial dairy ration, fed at a daily rate of 1 pound for every $2\frac{1}{2}$ to 3 pounds (depending on breed size) of milk produced, plus 2 pounds of good legume-grass hay for every 100 pounds of body weight. The dairy ration is composed of coarse grains with a 16-, 18-, or 20-percent protein level. Generally a 16-percent formula is needed, but if you are feeding a very poor quality late-cut hay, you may need the coarse 18- to 20-percent feed.

Using this feeding plan, a daily ration for a 900-pound lactating Jersey cow producing 34 pounds of milk is 18 pounds of legume-grass hay and 13.7 pounds of dairy ration.

A cow needs to have her milk supply dried off for six to eight weeks before calving. To dry her off, cut her grain ration down and stop milking. If her udder fills very full with milk, milk her out again, but only milk again as it becomes necessary. During the dry period the hay rations are the same, but feed only two to four pounds of grain per day, or up to six pounds per day if your cow is very thin. If she is on good pasture, do not feed her any grain during the dry period.

Once the calf is born, take it from its mother when it is a few hours old. For the first three days after calving, the cow produces a milk called colostrum, a thick, sticky liquid rich in nutrients necessary for the calf's health. Once the colostrum is gone, start using the milk for yourself.

Calves are fed warm milk twice a day, as much as 10 percent of the calf's body weight per day. Jersey milk is so rich that it is advisable to add one part warm water (101°F; 38°C) to three parts warm milk to prevent scours. Use a calf nipple and bottle feed, or teach the calf to drink from a pail by letting it suck your fingers after you've dipped them in the milk, then lowering them closer to the pail with each feeding until the calf is drinking from the pail.

If you are going to use the calf for veal, feed only milk, gradually increasing the amount according to increased body weight until the calf is eight to ten weeks old. If you are going to use it as a dairy animal, feed milk for four to eight weeks and begin a calf-starter grain and hay as soon as the calf will eat it. The calf should be ready for a calf-starter grain in one to two weeks and hay by the second week. When the calf is eating from $1\frac{1}{2}$ to 3 pounds of grain, wean it gradually from milk to water. After four to six months, calves can be fed the same grain mixture as mature cows. Calves are so susceptible to scours during their first month that you should seek the advice of your county agent in raising your calf. One lost calf in a large herd of cattle is not an economic catastrophe, but in a single family the loss is very significant.

BREEDING

To keep your cow milking the most economically, have her bred a couple of months after her last freshening. About sixty to ninety days after she has had her calf, start watching for signs of the cow's being in heat. She may act very friendly toward you, even trying to mount you; she may bawl and perhaps show strings of clear mucus on her tail end. Cows come into heat on an average of every twenty-one days and stay in heat an average of eighteen hours; the best time to have her bred is from the tenth hour to a couple of hours after the heat period ends. Once she has been bred, watch to see if she comes in heat again in twenty-one days; if she does, have her rebred. Most cows will settle with one or two breedings. Heifers are usually bred for the first time at fifteen to eighteen months.

A cow may be bred directly by a bull or by artificial insemination. You do not want to own a dairy bull because of his disposition and economic considerations, so unless you have a neighbor with a bull, artificial insemination is the recommended method. Artificial insemination firms are located throughout the nation.

Artificial insemination is a relatively easy procedure that allows you to breed your cow with a bull that has the finest characteristics. It is best to choose a bull that is of the same breed as your cow or at least of similar size. A small Jersey cow, for example, is apt to have trouble calving if she has been bred to a Holstein or Brown Swiss bull.

The cow's gestation period is about nine months, allowing for individual and breed variations. The average number of days in a gestation period for dairy breeds are as follows:

CATTLE

Ayrshire	279 days
Brown Swiss	290 days
Guernsey	283 days
Holstein	279 days
Jersey	279 days
Milking Shorthorn	282 days

MILK

A cow is milked in the same manner as a goat. First clean her off by brushing her down and trimming any loose hair around her udder. Wash her udder with warm soapy water and a paper towel and then dry it. Set your milking stool on her right side and sit so you can put your left knee against her hind leg to keep her from stepping in the pail. Wait a minute or so after washing her udder, then milk her as you would a goat (see pages 58–61); however, there are two more teats. Most people milk the front two teats first and then the back two. There are milking machines, but with one cow to milk you would spend more time milking and cleaning your equipment than you would milking by hand.

Use a stainless-steel milking pail and immediately strain the milk into a second one. Seamless stainless-steel equipment is the only way to ensure sanitation; strainers and filter discs are available at feed stores. Cool the strained milk to about 50°F (10°C) within an hour after milking; dividing the milk into multiple prechilled containers helps. Cooling slows down the growth of bacteria that cause souring and undesirable flavors.

The greatest level of milk production, about 40 percent of the yearly production, takes place in the first four months after freshening. Plan your cow's freshening date to fit into your schedule. Most homesteaders are busiest in the spring and summer of the year, so a fall freshening would be desirable. If your cow freshens in September or October you can have your harvest in and processed before the "white tide" hits. A calf born at that time of year will be well on its way before the cold weather sets in, and your cow will be dry during a time of year when it is cheaper to buy milk at the store. Hard cheeses also seem to be more successfully made during the cooler months.

COSTS

The cost of maintaining a dairy cow depends on the amount of pasture you have, the breed of cow, and the amount of milk she provides. To help give an idea of the expenses you would incur, we have used an example of a 900-pound Jersey cow producing an average of thirty-four pounds of milk a day in a 305-day lactation period. We assume access to a one- to three-acre pasture that would provide roughage for five months, and a purchase price of $750 for the cow just after her second lactation, which would make her a little over three years old; we divided the cost by nine, assuming she would produce until she was twelve. We further assume the successful raising of her calf as a vealer, but we encourage a prospective cow owner to view the calf as an extra dividend rather than dependable income.

COST OF RAISING A MILK COW

Purchase Price of Cow	
$750 ÷ 9	$ 83.33
Feed Costs	
Dairy ration (305 days) 4,148 lb. @ $11.45/cwt	475.36
Grain rations (dry period) 120 lb. @ $11.45/cwt	13.74
Hay (7 months) 3,906 lb. @ $81.00 per ton	156.24
Breeding Fee	
Artificial insemination	15.00
Bedding	
Low quality hay and/or straw	15.00
Total Costs:	$758.67

CATTLE

Product

Milk	
Lactation period of 305 days averaging 34 lb. of milk per day	10,370 lb.
1 calf, 100 lb. at birth, 200 lb. at 10 weeks, fed milk at rate of 10% of body weight per day for 10 weeks	– 1,033 lb.
Milk for owner's consumption:	9,337 lb.
Product Value	
Sale of calf, 200 lb. @ 75¢ per lb.	$150
Cost of milk per pound (total costs less sale of calf)	14¢ per qt.

Excess milk can be sold as raw milk to neighbors and friends, which reduces your overall cost per quart, or it can be converted to butter and cheese. Remember, however, that no labor costs are figured into the table.

MAINTENANCE

Cows are relatively easy to maintain in good condition. Adequate, sufficient, and sanitary food, water, and shelter are the basic sources of good health and condition.

The brushing down of a dairy cow before milking should keep the coat clean and help control lice. Cows can be washed with warm soapy water, rinsed, and the water scraped off with a water scraper or the backside of a comb. Most dairy cows enjoy getting a bath.

Proper hoof care is important, particularly when you consider how much weight those hoofs are supporting. There is a hoof trimmer made especially for cattle, designed so you can trim the hoofs without lifting the foot. Trim the excess hoof, but try to avoid trimming back so far that you cut the quick (exposed flesh).

A calf should be dehorned when a few days old. The easiest and safest means of dehorning is to use an electric dehorning iron. The procedure is the same as for disbudding goats (see page 77). It is not necessary to dehorn or castrate vealers.

If you buy your beef cattle as feeders and raise your dairy calves as vealers, you should not have to castrate any cattle, but for the record it is done with the aid of an emasculator. If you have a bull to castrate, have somebody (such as a veterinarian or an experienced cattleman) help you the first time; you do not want to slip and discover a few months later that your steer is really a bull.

Periodically you may have a need to restrain your cow (to give inoculations, check hoofs); to do so safely and successfully requires a bit of forethought. A calf may be restrained with relative ease by straddling it, but larger cows are more difficult.

A dairy cow may be restrained by stanchioning her and attaching a rope to her halter or nose clamp. Tie the rope so her head is pulled through the stanchion as far as possible and over to one side. To restrain the back end of a cow, grab the tail close to the body and pull it straight up and over the back. It is impossible for her to kick you when her tail is in this position.

Restraining a larger beef animal requires a chute. To be effective, a chute must be wide enough at the open end to get the animal to go in, and narrow enough at the closed end to keep the animal from turning around. A good stanchion at the narrow end will hold the animal once you get it down that far. There are commercial squeeze chutes, but they are too expensive for a single operation; you can design your own chute based on the commercial one.

MEDICAL CONCERNS

It is much easier to prevent medical problems than to treat an animal once it has become diseased. Proper diet and sanitation measures go a long way toward maintaining healthy animals. A cow's normal temperature is 101°F (38°C).

When you buy a cow be sure and ask what, if any, inoculations have been given. There are a number of vaccines available for cattle to prevent or reduce the chances of such diseases as anthrax, blackleg, brucellosis, hemorrhagic edema, leptospirosis, mastitis, and tetanus. Check with your county extension agent or veterinarian to see what inoculations are recommended for your area.

There are some tests that should be taken on your dairy cow every so often. The California Mastitis Test kit provides an economical and simple system of testing for this disease. It is refined enough to detect mastitis in its early stages, when it is easier to treat. Brucellosis (Bang's disease) and tuberculosis tests should be given every couple of years or as often as required by law. Brucellosis and tuberculosis are transmittable to humans, so you definitely do not want to neglect this chore. Both beef and dairy cattle can be affected by brucellosis and tuberculosis.

Blackleg is a disease caused by the ingestion of organisms living in the soil; it primarily affects cattle from six to eighteen months old. The symptoms are lameness, depression, and fever, and death usually occurs within forty-eight hours of the onset of symptoms. There is no treatment, but there is a vaccine available for young calves.

Bloat is caused by overeating, particularly on lush, high-quality pastures. The first symptom is a distention of the left side, which may be severe enough to raise the normal top line in that area. Bloat is serious and can be fatal, so call the veterinarian if it occurs. Administer up to a pint of defoaming agent such as peanut oil or mineral oil with a drench or stomach tube while you are waiting for the doctor to arrive. To help prevent bloat, keep feed barrels covered and slowly adjust cattle to a lush pasture by feeding about ten pounds of hay to the animal before turning it out on the pasture each day for a week or so.

Foot rot is caused by a fungus. The first sign of foot rot is usually lameness. Check the foot for signs of pus and swelling. Treatment includes cleaning the hoof and keeping it dry, and treating with a broad-spectrum antibiotic. The diseased tissue must be pared off the hoof with a knife. Confine the animal in a clean dry stall for seven to ten days. Preventive steps are to avoid muddy areas in the pasture and rough walk areas that contribute to hoof damage.

Johne's disase is caused by a bacteria and is a recurrent fetid diarrhea affecting primarily females from two to seven years old, most frequently after freshening. Any animal suspected of this disease should be isolated and tested. The disease is transmitted from infected animals and there is no treatment.

Ketosis is most often seen in high-producing cows just after calving. The symptoms include loss of appetite, drop in milk production, and depression. If you suspect ketosis, contact your veterinarian. Some homesteaders add a heavy concentration of molasses to the grain ration for a few weeks before and after calving as a preventive measure against ketosis.

Lice are small gray external parasites. The first sign of lice may be the cow's rubbing itself against anything handy. There are quite a few dusting powders available at farm-supply stores. Read the directions on the label and be sure to use only a powder safe for lactating animals on your dairy cow.

Mastitis is an inflammation of the mammary gland caused by various bacteria. There are acute, chronic, and gangrenous forms of mastitis. Symptoms include swelling of the udder, ropy milk, or a watery fluid in the milk. Gangrenous mastitis is indicated by a cold bluish coloration of the teats and bag. Mastitis can also occur in dry cows. Do not drink milk from a cow suspected of having mastitis. If you have a mastitis-testing kit, run a test on the milk. If not, contact your veterinarian. Mastitis can be treated with antibiotics.

Milk fever is caused by a calcium deficiency. It is seen generally only in cows with high milk production and usually just after calving. The first symptom may be a staggering walk. Paralysis will advance until the cow is unable to stand, and she usually holds her head at an odd angle. The cow's temperature will be subnormal. Milk fever is fatal if treatment is not started in time; call your veterinarian.

Pneumonia can be caused by viruses or bacteria. The symptoms are rapid breathing, a drop in milk production, a slightly elevated temperature, and perhaps a dry cough. Pneumonia can be treated with a broad-spectrum antibiotic. If you suspect pneumonia, isolate the animal and call your veterinarian. Remember to have a well-ventilated and cool barn in the winter to help prevent this disease.

Scours in calves are usually caused by overeating, and it can be fatal. If your calf's stools become liquidy, cut the amount of milk fed in half at the next feeding. If the situation does not improve, feed only water and call your veterinarian. When you calf does improve, gradually increase the amount of milk; never feed a calf more milk than 10 percent of its body weight.

Tetanus organisms grow in the absence of oxygen, so puncture-type wounds are your concern. You may have your cow given a tetanus vaccine; whether it has had a vaccine or not, have a antitoxin injection given following a puncture wound.

Udder edema is caused by an excessive blood flow to the udder and usually occurs in cows that have recently freshened. The bag will feel like dough and be swollen. Hot compresses and udder liniment will often alleviate the problem; if not, contact your veterinarian.

Worms of various types can invade cattle. Take a stool sample to your veterinarian periodically to determine what type of worms your cow has and treat accordingly. Periodic pasture rotation helps to keep down parasite ingestion.

A family cow or steer has a place on a homestead if you are willing to invest the time and effort and have pasture area available. The economics of raising a beef or dairy cow (or both) is certainly sound. A major factor in deciding whether a cow is for your homestead is your commitment to homesteading as a way of life. You must milk a cow twice a day for approximately 305 days a year, and a cow cannot be boarded out as easily as a goat. But a dairy cow will give you a great deal in return for your effort, and there is something extra special about owning a family cow.

8

GEESE, DUCKS, AND OTHER FOWL

ONE OF THE NICEST customs of the homesteading life is the practice of trading in kind with other homesteaders. You might, for example, help hay for a day in exchange either for hay or for a day of your friend's time during logging season. The custom is a money saver as well as fun. Our first flock of ducks and geese came to us in payment for disbudding some kids. I had read several books that referred to geese's "watchdog" ability, and had read an article that said a couple of full-grown geese could hospitalize a man by biting and beating with their wings. These bits of information had done a lot to alter my image of dear old Mother Goose, so we were a little cautious about the pair of white Embdens.

I have since learned that geese are not mean by nature. True, it is wise to be careful around them, especially when there are goslings (young) present, but that is because they are protective. Occasionally you will run into a mean goose, usually one that has been handled a lot as a gosling; this seems to diminish its natural hesitancy around people. If you are approached by a mean goose, do not run from it or you will probably get pinched. Just stand still and when the goose gets close enough, grab it by its neck and lift it off the ground for a minute or so. From that time on, that goose will be far more respectful of you.

Both geese and ducks are valuable to the homesteader. Aside from being a good source of eggs and high-quality down, they will earn their keep by alerting you to any unwanted marauders. They also help eliminate such garden pests as Japanese beetles and slugs. Geese and ducks are also an inexpensive source of delicious meat. One Muscovy duck hatching eighteen eggs can supply you with 110 pounds of meat at very little cost.

Geese and ducks are hardy, not prone to diseases or parasites, and require very little shelter. They do not have to have a pond or stream and can subsist on pasture for a good part of the year.

BREEDS OF GEESE

Geese have been domesticated for about three thousand years, longer than any other fowl. Its intelligence and stately carriage have certainly played a part in its popularity.

A group of geese is called a gaggle; the adult male is a gander, the adult female is a goose, and the young are goslings. The nest of eggs is referred to as a clutch.

There are about fifteen breeds of domesticated geese; those most commonly found on homesteads are the African, Canada, China, Embden, and Toulouse. Other breeds you might find are the Buff, Egyptian, Pilgrim, Pomeranian, Roman Tufted, Russian Fighting, and Sebastopol.

The African adult gander weighs about twenty pounds, the goose about eighteen pounds. The most common color is brown. These birds are predominantly light brown with a dark brown stripe down the head, neck, and back. Their wings are dark brown with a lighter shade as edging. The all-white variety of African is quite rare.

The bill and knob (the fleshy bulge at the base of the upper bill) of the brown variety are black, the eyes brown, and the shanks and feet orange (the white variety has an orange bill and knob as well as orange shanks and feet). Africans also sport a dewlap, which is skin that hangs down from the throat and upper part of the neck.

Canada geese, known to most of us incorrectly as Canadian geese, require a federal and sometimes a state certificate to be held by the owner stating that the geese were acquired from a breeder who has a federal and state permit to raise the breed. If you in turn want to sell or even give one of the birds away, you must have received the federal and state permits.

The beauty of the Canada makes the breed worth the bother of the permits. Their black heads and necks are set off by white cheek patches on the head. The back and breast are gray and the tail is black. Ganders weigh about twelve pounds, geese about ten pounds.

Chinas are probably the most common homestead breed. The China's popularity is due to its being the most prolific egg layer of the various breeds and to its foraging ability. They not only will forage large areas of feed, but they will eat crabgrass, ragweed, and pigweed, which makes them highly desirable creatures. They are, however, the noisiest of the goose family.

China geese come in two varieties, white and brown. Both varieties have orange feet and shanks. The white have orange bills and knobs, and the brown have a dark, slate-colored bill and knob. The adult gander weighs about twelve pounds, the goose about ten pounds.

China geese are the breed most commonly used as weeder geese. There has been a new interest among organic gardeners in using geese for weeding crops. Weeder geese have been used successfully in fields of strawberries, cotton, potatoes, and other crops. A gaggle of six to eight geese per acre can eliminate grass and weeds in the garden.

There is one major problem with weeder geese on the homestead: because the homestead is most likely a multicrop venture, the geese will eat young corn and other desirable plants. A good fencing system, however, permits the homesteader to use weeder geese.

Embdens are large geese; the adult gander weighs from twenty-six to thirty pounds or more, and the goose weighs about twenty pounds. These geese are pure white with blue eyes and orange feet and bills. Sometimes the young geese will have gray feathers in the wings or on the back.

Toulouse geese are regal in appearance. Adult males will weigh about twenty-six pounds and adult females about twenty pounds. The Toulouse comes in varying shades of gray with a light orange bill and dark orange feet. The dewlap is quite large and wrinkled.

BREEDS OF DUCKS

Ducks are also useful on the homestead; they provide good meat, eggs, and are very hardy and inexpensive to raise. They will help rid your garden of pests, and if you are fortunate enough to have a farm pond, they will help cut down on the algae. With a pond, of course, you must limit the number of ducks to fit the size of the pond. Besides all of these fine qualities, ducks are easy to raise and just plain fun to watch.

There are a number of duck breeds, and all except the Muscovy were developed from the Mallard, a wild duck. The breeds of ducks most commonly found on a homestead are the Indian Runner, Khaki Campbell, Pekin, Rouen, and Muscovy. There is some discussion as to whether the Muscovy should be classified as a duck or a goose. Since another name for Muscovy is Pato, which means duck in Spanish, we will include the Muscovy in the duck section.

An adult male is a drake, an adult female is a duck, and duckling is the term used for newly hatched ducks that are not completely feathered. A drakelet and a ducklet are respectively a male and female in their first year.

The Indian Runner is generally considered the Leghorn of the duck world—that is, the most prolific egg layer. The eggs are white or whitish in color. Excellent foragers and weighing only about $4\frac{1}{2}$ pounds, Indian Runners are economical to raise. Their meat is tasty, though not abundant. The Indian Runner has a tall, narrow body. There are a number of color varieties, the most common being all-white, fawn and white, and pencilled.

Khaki Campbells are an excellent breed, especially if you are interested in a dual-purpose duck (egg production and meat). The Khaki Campbell is a close rival of the Indian Runner in egg production and is also a hardy lightweight bird, adult drakes and ducks weighing about $4\frac{1}{2}$ pounds. The Khaki Campbells are excellent foragers and therefore economical to raise. Khaki Campbells closely resemble Indian Runners except in color; the Campbell male has a khaki brown body and a brownish bronze to greenish head, lower back, and top tail feathers. The females are khaki everywhere but the head, where they are medium to seal brown.

Pekin ducks are what most people think of as a duck: Donald Duck is a Pekin, as in the Long Island duck (of dining fame). The Pekin is a good egg layer and excellent for meat, as the young grow very rapidly. The adult drake weighs about nine pounds, the duck about eight pounds. The Pekin are white with orange bills and feet. We have not had much success with the Pekin ducks as mothers, but perhaps this is just characteristic of our particular Pekins.

Rouens are beautiful ducks and though larger than the Mallard, they are very similar in color, with brown feathers and, on males, a green head and neck. The adult drakes weigh about nine pounds and the ducks about eight pounds. The Rouen is a good layer and the meat is excellent. The Rouen is a little harder to dress out because of the dark feathers. Our Rouens have been the least friendly of our various duck breeds, but their young have always sold the fastest because of their beautiful coloration.

Muscovies, in our opinion, are the best multipurpose breed of duck. They are not only good egg layers, but their hatchability rate is high. The ducks are good mothers. Their large size makes them excellent meat birds. Muscovies are large, with drakes averaging twelve to fourteen pounds and ducks seven to nine pounds. The fall dispatching of spring ducklets and drakelets will yield at least a four- to six-pound carcass. The Muscovy looks more like a goose than a duck; its body is long and broad and nearly horizontal in carriage. There are white, blue, chocolate, and buff varieties. They are all good ducks, but the white is easier to dress out because of its lighter plumage. Muscovies have rough, fleshy protuberances called caruncles on their faces; the drake's caruncling is very pronounced.

Muscovies have two other unique characteristics. First, they are known as quackless ducks; the noise they make sounds like a gentle hissing or heavy breathing. And second, Muscovies have less oil in their feathers than other ducks and therefore do not take to water as other ducks do. They can swim but are not too fond of it. When they are in molt (the gradual process of feather loss and replacement), the amount of oil is so low that they could easily drown if in water too long. Both ducks and geese molt, usually in the fall.

SHELTER AND EQUIPMENT

Geese and ducks need very little in the way of shelter. Their basic need is to be kept out of the wind, as dry as possible, and clean. A three-sided shed open to the south is ideal. You may provide a platform with straw or hay as bedding if you live in a very cold or wet climate. If you are sheltering both geese and ducks or different breeds, provide a large enough space or separate shelters for each group. Enclosed shelters are not very successful, as ducks and geese easily become panicked if they are confined and a strange noise or movement makes them nervous.

In the spring, supply some nesting areas for your geese and ducks. We have found old bureau drawers, old tires, and doghouses lined with straw to be pleasing to ducks and geese. However, do not be disappointed if they bypass your efforts and establish their own spot. They like to nest in a relatively secluded spot that includes privacy from other sitting fowl, even of the same species.

If you need to fence in your fowl, have the penned areas on a slope and place the feed and water at the highest spot. A portable pen is excellent if you have enough land on which to move it around. The best system is to fence in your young plants and let your fowl roam so that you can rely on pasturing for free feed.

If you do not fence in your fowl, you must pen or cage them for three to five days when you first bring them to your homestead. This way they learn where home is and where food and water are obtained. If you do not do this, they will probably take off. If you hope to mate a certain pair, cage them together during this time, even if you already own one or two fowl, to establish them as a unit.

It is possible to use electric fencing for geese and ducks. Run the bottom wire about six inches off the ground and the top wire at least chest high to your largest bird. If you use electric wire, check to make sure the weeds do not short out the wire.

Geese and ducks need lots of fresh clean water to drink at all times. The water containers should be deep enough for even the geese to be able to submerge their heads totally. The ducks and geese will undoubtedly do some of their bathing in their drinking water if the containers are wide enough to accommodate them; small containers will assist you in keeping the drinking water clean.

Geese and ducks do not need a pond or brook on the premises. They do enjoy swimming and need to bathe, but a child's pool or a large bucket will suffice. If you use a plastic pool, surround it with gravel to keep the birds from tracking in mud. They will love the pool and you will get more than your investment back in your enjoyment in watching them.

FEEDING

Geese and ducks can subsist on foraging if your pasture is ample enough. Good pasture consists of grasses, clover, and weeds. They can graze a yard, field, river or pond bank, or marshland.

To feed fowl that are penned up you need to provide mashes, pellets, crumbles, or grain. Feed your geese twice a day and ducks once a day. Do the single or heaviest feeding at night when supplementing pasture so as not to discourage foraging. Feed them only what they can clean up in fifteen minutes.

Mashes, pellets, and crumbles are commercially prepared food, ensuring a balanced diet for your fowl. Choose an egg-layer mash without medication, for it does not benefit waterfowl. Follow the manufacturer's directions for amounts, which will vary according to the age of the fowl and the amount of foraging they do.

Geese and ducks are far more efficient feed:meat converters than chickens. It takes an average of 22 pounds of feed to produce a 6½-pound ducks, and approximately 42 pounds of feed to produce a 12-pound goose, a conversion ratio of less than 3:1.

Geese and ducks love corn—on the cob, whole, or cracked. Corn is especially valuable in the cold months, but use it sparingly in the warm months. Bread can be used to supplement their diets; it tends to be about 16 percent protein and they love it. Old bread can be purchased for pennies a loaf from any nearby bakery.

Discarded produce such as lettuce, cabbage, and fruits will be much appreciated by your fowl, especially in the winter. In the summer, if you are penning up your fowl, do not overlook grass cuttings as a food source.

Geese and ducks need coarse sand for grit and a good source of calcium, such as oyster shells; you can also smash up eggshells and feed them as a calcium source. If you feed a commercial feed you do not need a calcium supplement.

COSTS

It is very difficult to determine costs of raising waterfowl because of the differences in foraging potential among homesteads. We can give some rough and conservative estimates, figuring the use of poor to average foraging potential. This estimate covers the cost of a high-level usage of grain and concentrates in the coldest areas of the country.

ANNUAL COST FOR RAISING DUCKS

Breeding	
Initial cost of 2 breeding-stock ducks @ \$2.50 per duck	\$ 5.00
Feed for 6 months a year @ $2\frac{1}{2}$ lb. per week, 130 lb. @ 9¢ per pound	11.70
Feed for rest of year (assumes forage of half of diet), 651 lbs. @ $1\frac{1}{4}$ lb. per week	5.85
Ducklings	
Assumes 8 offspring maintained for 12 weeks, half of diet by foraging; 11 lb. of feed per duckling @ 9¢ per lb.	\$ 7.92
Total Cost:	\$30.47
Product Value	
Assumes 8 young ducks, dressed weight of 4 lb. each at 12 weeks; 32 lb. of meat @ 95¢ per lb.	95¢

The foraging ability of ducks and geese is so good that you probably can raise a considerable flock of them for the cost of one bag of feed per year, bringing the cost per pound down to just pennies.

BREEDING

The primary breeding season for geese and ducks is in the early spring. Most ducks will set a second clutch in a year after they have finished raising their first brood.

There are four breeding methods: inbreeding, the breeding of close relatives (father-daughter, brother-sister); linebreeding, the breeding of less closely related birds (aunt-nephew); outcrossing, the breeding of two fowl of the same breed that are not related; and crossbreeding, the breeding of two birds of different breeds. Linebreeding is the best choice for improving your line, although many homesteaders successfully use inbreeding.

If you have more than one breed of duck or geese, you will undoubtedly have some crossbreeding unless you pen up your breeding stock for the duration. Crossbreeding does not take place between geese and ducks, but rather between breeds of ducks and breeds of geese.

A major disadvantage of crossbreeding is that you do not know what the product will be, or even if there will be a product. Muscovies, like most ducks, will readily indulge in sexual activity with other duck breeds. The chances of the resulting egg's being fertile are small; if an offspring is produced, the chances are that it will be sterile.

Ducks are sexually mature at about six months for the light breeds and seven months for the heavier ones. Drakes of all breeds do better during their second breeding season. A duck's productivity starts to decline significantly after three to five years. A drake maintains a high degree of fertility for eight to ten years. The light-breed drakes can service four to five ducks; the heavier-breed drakes, two to four ducks. Muscovies can service four to six ducks.

Geese and ganders reach sexual maturity during their second breeding season, with the exception of the Canada, which may not reach sexual maturity until her third or fourth year. Some geese will lay eggs during their first year, but generally the results are inferior.

Ganders are fertile for about ten to twelve years, but a goose can go on much longer. The gander:goose ratio is 1:2 or 1:3 for the

Africans, Embden, and Toulouse; Canada geese mate only in pairs; and the China gander services an impressive harem of four to six.

For any breeding to be successful, there must be at least one male in the crowd. You may easily sex (determine the gender of) Rouens, because male and female have distinctive coloring. To sex duck breeds without color differentiation, look for the curled sex feathers on the tails of the adult males. They are quite obvious even from a distance. Muscovies are the exception to this rule, but their heads make distinguishing their gender easy. Adult geese are most easily sexed by the difference in their sizes, males being larger.

Young ducks and geese may be sexed by checking the sex vent under the tail. Put a little petroleum jelly on your index finger and insert your finger about an inch into the vent. Move your finger around gently a few times to relax the sphincter muscle, then apply pressure around the vent to expose the sex organs. The male organ is corkscrewlike, unlike the smooth and less prominant female genital. Sexing fowl this way is not easy for the inexperienced. It is useful to first sex birds of a known gender so you have a clear idea of what you are looking for.

The final proof of your ability to sex fowl will come in the spring: the birds on top are the males of the species.

EGGS

Ducks usually lay their eggs by midmorning. Geese are more erratic, so you need to check several times through the day if the weather is cold.

Duck and goose eggs are nutritious and taste delicious. Duck eggs are considered to be more nutritious than chicken eggs as they are higher in protein content, though they are also higher in cholesterol. One hard-boiled goose egg makes a delicious egg salad sufficient for four hungry people.

Goose eggs are in demand by people who make decorated eggshells. In fact, selling goose eggs to craft centers can easily pay for a couple of big bags of grain a year.

GEESE, DUCKS, AND OTHER FOWL

Geese and ducks that live in the colder regions of the country start laying in the spring, before the danger of frost is past. You should therefore pick up the eggs to prevent their freezing. Watch to see where your geese and ducks nest so you can find the eggs quickly. Generally they will not switch their nests if they don't see you taking the eggs.

Mark the eggs you have gathered to hatch with both the date and the type of egg. The hatchability of stored eggs begins to decrease after ten to fourteen days. The eggs should be kept at a temperature of 35°F to 50°F (2°C to 10°C) and in a moist environment. This can be accomplished by draping a damp towel over the top of the egg container and keeping the towel damp. The eggs are best kept in a carton or other container that permits them to be stored without being stacked on top of each other. Turn the eggs once each day on the axis.

Incubation can be accomplished by the natural mother, surrogate mother, or a machine. The best method depends on what you plan to do with the eggs, the skill of the natural or surrogate mother, and your time schedule.

When the weather is past danger of a severe frost, set the eggs in the nest of the brooder goose or duck. Be sure she has already laid an egg in the nest the day you plan to add the ones you have stored; she will then accept all the eggs. Once she has a clutch to set, she will stop laying eggs.

Any good brooder goose or duck will set any other goose's or duck's eggs, and so will a good brooder chicken. A large hen can set about five goose eggs or nine duck eggs. Make sure that the hen covers all of the eggs completely. If you are using a hen, you must mist goose and duck eggs with warm water once a day to ensure proper humidity (a goose's or duck's daily swim or bath does this naturally for her own eggs). Also, you must turn the goose eggs several times a day, as they are too heavy for a hen to turn. A hen that is broody enough to put up with all of this is indeed a real treasure and should be appreciated greatly. Either a Cochin or Bantam hen is a good choice for a surrogate waterfowl mother.

If you use an incubator, make sure it is large enough to accom-

Incubation Periods for Homestead Fowl

Bird	Egg-laying Ability	Incubation Period (in days)
African	moderate	30–34
Canada	low	28–30
China	excellent	28–30
Embden	moderate	30–34
Toulouse	good	30–34
Indian Runner	excellent	28
Khaki Campbell	excellent	28
Pekin	good	28
Rouen	moderate	28
Muscovy	moderate	33–35
Turkey	low	28
Guinea fowl	good	28
Chickens	good-excellent	21

modate waterfowl eggs and then follow the manufacturer's directions. It has been reported that waterfowl eggs hatch better in still-air than in forced-air incubators.

The hatching ability of geese and ducks varies a bit from breed to breed and from brooder to brooder. If you are letting the geese and ducks hatch their own eggs, remember that they will come off their nests for food and drink usually only once a day. Therefore, make sure there is a ready source of food and water for them when they need it.

GOSLINGS AND DUCKLINGS

The young do not require food or water for the first twenty-four hours after hatching. This allows the mother to remain on the nest for a longer period of time to allow the rest of the eggs to hatch. Remember, the eggs are not all going to hatch at the same time. When the mother needs to, she will leave the nest. If you are game, check the nest for remaining eggs after the mother and the young leave. If you find some left, keep them warm and they may hatch out. Do not become overanxious, however, and try to help the bird hatch by removing the shell.

Geese and ducks are generally good parents to their own young. The more active breeds may tend to walk their young too rigorously in the first week or so. Penning may be a good idea if you have a large area where they normally roam. A second potential problem arises if you have an insufficient area for the goslings or ducklings. The better the mother a goose or duck is to her own young, the more vicious she tends to be toward another mother's young who try to join her brood.

Drakes are not interested in helping raise ducklings. In fact, one occasionally runs across a drake that should be separated from the ducklings until they are older for the ducklings' protection. On the other hand, ganders make especially good parents. They are highly protective and proud of their young. Ganders have been known to stand guard over an unhatched egg or two in a nest long deserted by the goose. It is therefore advisable to pick up any unhatched goose eggs.

If you are using a brooder hen, be sure to confine her for three to four weeks, or she would undoubtedly exhaust the young goslings or ducklings. If you are confining the young, keep the area clean and dry. Use a commercial absorptive bedding or peat moss.

If you have elected to incubate the eggs, you will need a brooder once they hatch. The ideal brooder provides a constant temperature of 88° to 90°F (30° to 32°C) for the first three days, with a gradual reduction to 80°F (27°C) by eleven days after hatching. By three weeks of age, the eggs should be adjusted to a temperature of 70° to 75°F (21° to 24°C) and at four weeks, a temperature of 70°F (21°C) is sufficient. A big box with bedding and a 100-watt lightbulb will also work. (For additional information on brooders, see Chapter 4, pages 83–107.)

Goslings and ducklings need a constant supply of clean water; the waterers should be deep enough for them to submerge their heads. Avoid providing excessive amounts of water to the young. They will swim, but they do not have the feathers or oil to keep from getting chilled. The young can even drown if their down gets too wet and heavy, keeping them from getting out of the container.

Goslings and ducklings will eat grass and insects, and commercial feed if supplied; commercial feed helps ensure a balanced and sufficient diet.

BUTCHERING

Ducks are ready to butcher when they are about ten to sixteen weeks old; geese are ready at about sixteen to twenty weeks. Our experience has been that the feathers are easier to pluck if you wait until the outside of these periods.

Waterfowl are butchered in about the same manner as chickens (see pages 97–98). Hang the bird by its feet at your shoulder height and insert a sharp pointed knife through the wedge-shaped opening in the roof of the mouth. Cut from front to back with a twisting motion until blood starts dripping; then draw the knife across from that incision to the corner of the mouth. It will take a little pressure. If you do not use a killing funnel, hook a wire with a weight on it through the lower beak to keep the fowl from spraying you with blood.

Once drained, dip the bird in hot water (140°F; 60°C). Agitate the bird for four minutes, then test to see if the feathers are loosened. When the feathers are ready, hang the scalded bird at shoulder height and start picking off the wing and tail feathers. Use a strong twisting motion. If the feathers are hard to remove, dip the bird back in the hot water for half a minute or so and then try again.

Use paraffin wax to remove the pinfeathers. First dry the bird off while melting some wax in the hot water. Then dip the bird into the pail several times. The wax will adhere to the carcass. When the wax cools, peel it off and the remaining feathers will come with it. You may want to singe off the hair which remains after plucking; use a propane torch or tightly rolled newspaper. Remove the head and feet when the bird is all plucked. Cool the carcass overnight before cleaning.

To clean the bird, make a lengthwise incision almost to the breastbone. Cut out the vent, loosen and remove the intestines, giblets, kidneys, and lungs. Next, slit the neck skin down the back and remove the crop and windpipe from the neck opening. Cut out the oil sack above the tail, being careful not to break it. Rinse the bird both inside and out and it is ready to freeze or cook.

MAINTENANCE

Geese and ducks require very little maintenance. A crude shelter, food, and water are the main requirements. In fact, they are so easy that the homesteader's biggest problem is to curb the tendency to collect too many.

Most geese and ducks do not fly enough to present a problem. Two exceptions, however, are Canadas and Muscovies. Canadas should be pinioned when they are first born or they might migrate when they become fullgrown. Muscovies are good flyers, but they are not apt to fly away, and their flying ability aids them against dogs and other enemies. Pinioning (wing clipping) is described in Chapter 4, page 104.

Down can be plucked any time after the weather gets warm. Plucking down off a live goose or duck is painful to the bird and really unnecessary. If you wait until you butcher the fatted fowl for dinner, you can collect all the down you want. After you have plucked the bird, put the body feathers and down in an onion sack and hang it up so the feathers can dry. Every day, fluff the bag as you would a pillow to bring the wet feathers to the outside. When the feathers are completely dry, you can use them to make a vest or pillow.

MEDICAL CONCERNS

Geese and ducks are highly resistant to disease as long as they are properly taken care of with a balanced and sufficient diet, fresh water, enough room, and clean quarters. They are subject to some ailments, however.

Bumblefoot is the development of a hard, warty cushion on the feet of fowl. The bird will limp noticeably when afflicted. The problem may result from a bruise or a staphylococcus infection. To treat, put the bird in a pen with lots of clean bedding and apply an antibacterial ointment to the foot until the cushion softens, which takes a few days. When it softens, you may be able to remove a hard core or you might see yellow pus. Clean the area and continue to apply ointment until the bird regains full use of the foot. If pus is evident, add oxytetracycline to the bird's drinking water, following directions on the package; it is available from feed stores, veterinary supply stores, or from your veterinarian.

Geese with knobs should be provided with shelter if you live in areas with severe winters, as their knobs are subject to frostbite. If frostbite occurs, treat with cold packs and then apply a healing ointment or petroleum jelly. If the frostbite is mild, there may not be any permanent discoloration.

Phallus prostration is the failure of the gander's or drake's penis to retract. This condition may be treated by isolating the bird and coating the organ with an antibacterial ointment. If the condition occurs for no specific reason, it may well be a genetic factor and you would do well to separate the bird to avoid passing on the trait. It more frequently occurs in a drake that has too many ducks to service or a drake that is too young to service the appropriate number of ducks.

Staggers, as the name suggest, is a condition in which the bird, particularly a young one, staggers around. It is usually fatal. It is caused primarily by an insufficient amount of water, but can also be the result of a lack of shade during extremely hot days.

TURKEYS

It is not impossible to raise turkeys, but it can prove to be a challenge. Turkeys are more disease-prone than geese or ducks, are very easily frightened, and are unbelievably stupid. However, if you can keep them healthy, teach them to eat and drink, and get them under shelter whenever it rains, you can provide your family with a delicious homegrown turkey for Thanksgiving that will cost you very little money.

Turkeys are basically economical to raise. The feed:meat conversion ratio is 3:1, and they dress out at about 80 percent.

The Bronze turkey is the big beautiful bird we see pictures of around Thanksgiving. The disadvantage of this magnificent bird is that it is the hardest to dress out because of the dark feathers. The White Holland and the smaller Beltsville White are the two other common breeds.

Baby turkeys are called poults. You want poults that are from a hatchery that is U.S. pullorum clean. Get your poults after the warm weather has settled in and you can simplify your brooding period. Keep the poults in a brooder, if you have one, for a couple of weeks; a large packing box with a 60- to 100-watt bulb will do. Use peat moss or a commercial absorptive bedding.

Provide water in a chicken waterer. A few marbles in the water lip will encourage them to peck at the water and learn to drink. To be on the safe side, however, dip their beaks into the water several times the first day you have them.

You may feed in chicken feeders or a trough box. To get the birds to eat, use turkey starter mash with a little chick scratch on top of it. The chick scratch is coarser and seems to get them going. Once they are eating, you can eliminate the chick scratch. If they aren't eating by the second day, try sprinkling some of the feed on clean white paper to increase its visibility. If that fails, try stuffing a small bread crumb down them once or twice. As you can see, it would not be wise to start off with too many poults.

The best way to raise turkeys is in wire cages above the ground; each turkey requires five square feet of space. The floor screen should be made of half-inch-mesh heavy-gauge wire. Turkeys can be raised on the ground, but then you have to make sure they are in an area that is and has been chicken free (to avoid blackhead, a turkey disease); you also have to worry about them during wet weather. The turkeys will need a roof over at least half the cage to provide protection from both the sun and rain.

Commercial pellets provide an economical balanced diet for your birds. You may supplement their diet with greens from the garden. The greens, however, should only make up 25 to 40 percent of their daily diet. Turkeys like grasses of all types and most garden greens.

Turkeys are often finished (fattened up for slaughter); corn is the most common finishing grain. It is fed as the only or primary feed the last three to four weeks.

Turkeys are quite susceptible to diseases; the most common is blackhead, caused by a protozoan hosted by the small roundworm prevalent in chickens. Any chicken equipment you use for your turkeys should be disinfected thoroughly before use. A bird that has blackhead will look sick and have yellow droppings. There are medications to control blackhead, but you are much better off preventing it in the first place.

Pullorum should not be a problem if you buy from a hatchery that is U.S. pullorum clean. Due to the testing programs, pullorum is not prevalent today.

GUINEA FOWL

A definite favorite of homesteaders is guinea fowl. These noisy, cantankerous birds were semidomesticated by the ancient Greeks and still are more wild than tame. The reason homesteaders favor them is because they are fearless protectors of the farm against stray cats and dogs, incredibly effective against all insects (especially the potato bug), can produce up to 100 eggs a year, and are a real gourmet delicacy on the platter. The meat is very similar to pheasant in flavor, and I will wager has appeared in many a restaurant "under glass."

The three most popular varieties of guineas in this country are the Lavender, White, and Pearl. As guineas are primarily a meat bird, choose breeding stock with full, meaty breasts.

The guinea eggs can be eaten. They are about the size of Bantam chicken eggs, with extremely hard shells. Incidentally, guineas can cross with chickens, but the offspring will be sterile.

Adult guineas can be allowed to roam freely. They do not scratch as much as chickens do, so they are not as damaging in the garden. They prefer to roost in trees. Guineas will always come home at night to roost no matter where the day's travels have led them. At night, there will always be one guinea left to stand guard.

If they are allowed to roam, they will feed themselves whenever possible, but in the winter, offer them grain and turkey feed. They also should have oyster shells and grit, and fresh water should always be provided.

One of their most difficult habits is hiding their nests. They are also not always good mothers, so if you are counting on raising some of the young, you may want to hatch some of the eggs. Once you locate the nest, collect eggs for the incubator. Leave two eggs so the hen doesn't change her nesting spot. Be sure to mark the eggs you leave so you always gather fresh ones. Using this system, you can get one hen to lay many eggs. The hen will set on twelve to fifteen eggs.

The eggs you have collected can be eaten, incubated, or given to a broody Bantam hen to hatch. If you are going to use a chicken, keep the eggs at about 55°F (13°C) until you have enough eggs for the nest. If you are using a Bantam, twelve to fifteen eggs is about right.

Young guineas are called keets. Keets can be fed turkey feed or allowed to forage. Keets are rather prone to drowning, so whether you use a shallow pan or a regular chick waterer, put small stones in the container to make the water shallower.

Sexing guinea fowl is difficult. The cocks have a slightly heavier and coarser head and have wattles (fleshy folds of skin at the neck). The cock also has a one-syllable shriek, whereas the hen has both the one-syllable and a two-syllable shriek. The two-syllable shriek sounds a bit like the hen is saying "buckwheat."

The word *shriek* for the guinea fowl's call is not casually chosen. Guinea fowl should not be raised if you have close neighbors and want to stay on friendly terms with them. Guinea fowl can be unbelievably noisy. When our friend delivered our guineas, he had a dozen more in his truck. I asked if he were going to keep some of them, and he said, "No, I just have six acres and that's not enough for these birds."

Of all the advantages of geese and duck on a homestead, the greatest by far will be the sheer pleasure you will derive from just watching them. They are fun.

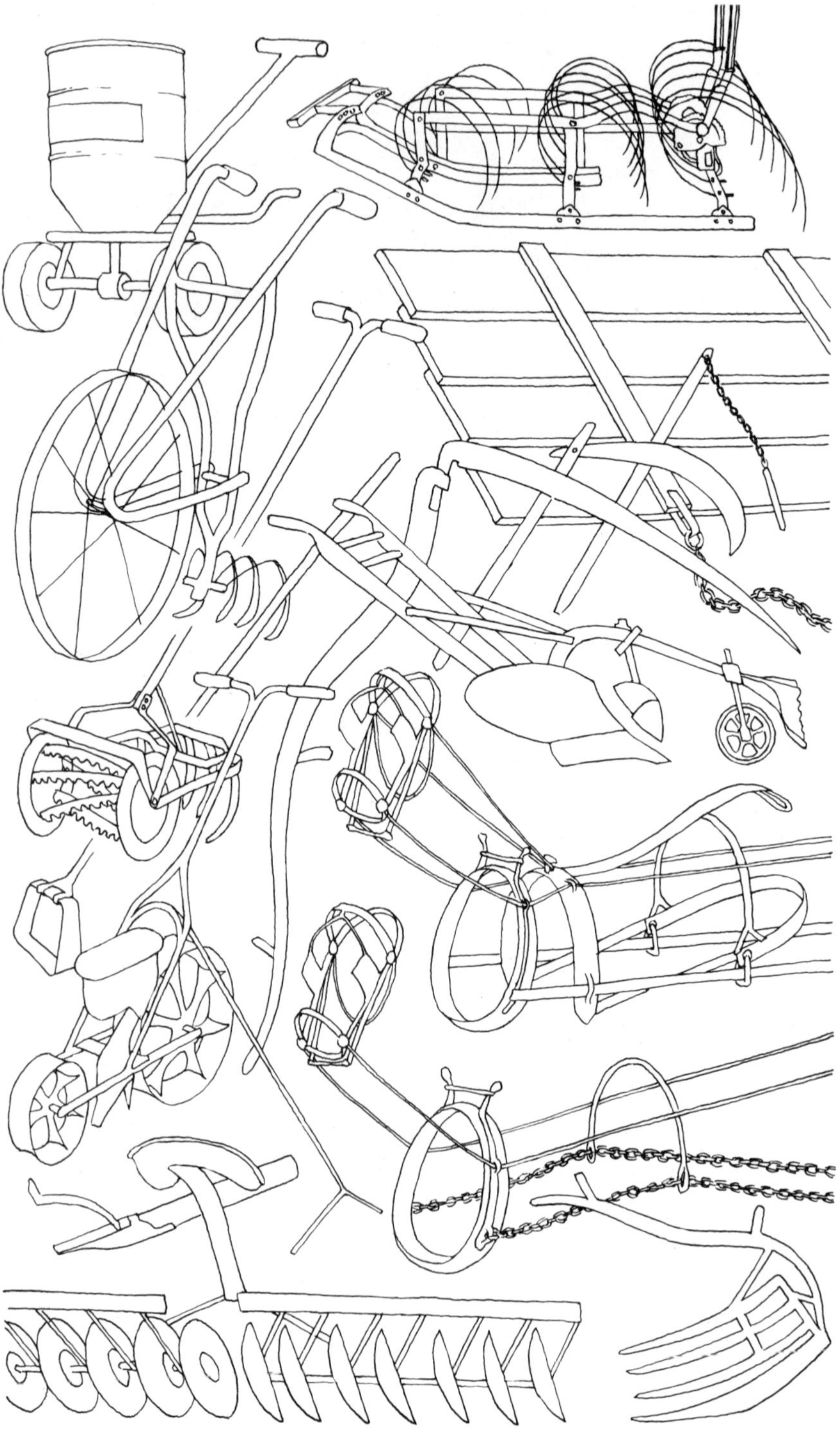

9

HOMESTEAD ENERGY AND TOOLS

HOMESTEADING requires a lot of human effort. The results of this effort can be increased by coupling people with other sources of energy such as animals, internal-combustion engines, and electric motors. Energy sources can accomplish specific tasks by attaching them to appropriate tools. Choosing the sources and tools for the broad range of homesteading tasks presents a great number of decisions.

For example, in breaking ground (turning up soil), the source of energy can be good old muscle power; one or more ponies, horses, or oxen; two-wheel tractors; two-wheel-drive, four-wheel tractors; or four-wheel-drive tractors. The power range of tractors is practically limitless, from 5 horsepower to more than 100 horsepower. Then there are tools: the homesteader can use a round or square-pointed shovel with a long or short handle, or a spading fork; in conjunction with a two-wheel tractor, the homesteader has to decide whether to use a moldboard plow or a front- or rear-mounted rototiller; with four-wheeled tractors, he or she has to decide whether to use a rototiller, moldboard plow, disc plow, or chisel plow. This chapter focuses on the options to consider in making these decisions. We have tailored the options for the homesteaders with up to five acres.

BACK AT THE FARM

When we started homesteading, our range of options was limited by our circumstances and conditions. After buying our homestead, we did not have much money to invest in equipment. When we first wished to break ground for a garden, all we had was a round-pointed long-handled shovel. Our garden spot had not been worked for ten years, so it had a heavy growth of weeds and some brush on it. Coming from the city, we were not physically fit enough to spade up this large area. The local rental shop only had front-end, five-horsepower rototillers, insufficient for a job of this magnitude. We knew of no one who did custom plowing, but we did know Bruno, who did custom rototilling with a large tractor-mounted tiller, and we decided that Bruno was our best bet. We have since learned that rototilling is an excellent way to break up old ground (land which is worked every year) and fit (prepare the seedbed) simultaneously. It is not a good way to break up new ground (ground which has not been worked for many years), especially new ground which has been permitted to go to weed for ten years.

Fitting the seedbed in this garden was very difficult with the steel-pointed rake because of the amount of organic trash on the surface. We also had a hand cultivator, but with the amount of trash and our poor physical condition, it was almost impossible to push. We did have two horses and a pony (better for homesteading if we had had no horses and a tractor, but horses were our primary reason for moving to the country). The pony was the only one of the three equines who knew how to drive. Therefore, Freddie became our primary source of energy for the first year. We hitched him to the hand cultivator with a light driving harness and prepared a beautiful seedbed.

The following year we negotiated with Lahneen to break the old ground and fit the seedbed. (You never break an old horse to work; you just negotiate an agreement.) The following year we expanded our crop land and Nappy broke up our new ground with his International Cub tractor and a moldboard plow. We fitted it with the two horses which were now working as a team. The next year we bought an old oil-smoking 9N Ford tractor with a mounted two-bottom moldboard plow. Now we can turn the whole homestead over in a half a day. In the years since, we have used the tractor, an old pickup, and horses on the jobs for which they are best suited. Some days we long for the romance and exercise of using the team exclusively; other days we covetously eye the shiny new small four-wheel-drive tractors and full line of mounted equipment.

HUMAN POWER

The basic homestead source of energy is human power. All members of the family can provide valuable energy. In addition to accomplishing tasks, expending human energy benefits the person by:

- Improving the efficiency and capacity of lungs, heart, and other organs
- Improving the strength, response speed, and endurance of important muscles
- Sharpening mental processes
- Releasing tension

It is possible for homesteaders to rely on human energy as the sole source of power. It requires the least amount of capital outlay, but demands the greatest amount of time from homesteaders. A person plus a tractor pulling a two-bottom plow can break hundreds of more units of ground than the same person using a shovel. The following variables must be considered in deciding to rely solely on human power:

- The total human energy available (number of people and relative strength of each person)
- Time available of each person (especially at the peak of homestead demands)
- Money available for capital investment in other sources of power and tools
- Size of homestead
- Nature of tasks
- Personal interest of family members

If you are a large family with older children who are too young to work off the homestead, do not have much money, live on a half-acre homestead, and rely primarily on vegetables for your diet, a horse or tractor may not have merit. On the other hand, if you are a couple with young children and both of you work off the homestead, commuting some distance, and wish to raise lots of animals on your five acres, you will probably wish to purchase a horse or two, or a tractor.

For most people, the variables will not be so clearly defined. The case may be more like the following: the husband (who is mechanically inclined) works long hours off the place. There is only one son, who is sixteen and would like a job off the homestead. There is one daughter who loves animals, especially horses. The homestead consists of three acres and the family are big meat eaters. In this case the decision is not so clear-cut. Chances are, this family will choose a combination of power sources, including human power, horsepower, and tractor power.

Whatever your circumstances, it is best to start with what you already have when determining your energy needs. You already have the potential of the family members; to this basic inventory, decide what power source, if any, to add. The major homestead tasks are as follows:

Outdoor Tasks	**Indoor Tasks**
Breaking ground	Separating cream
Fitting the seedbed	Churning butter
Planting	Cutting and grinding meat
Cultivating the plants	Grinding flour
Harvesting	Squeezing and straining vegetables
Processing the harvest	Chopping vegetables
Moving refuge	Slicing and pureeing vegetables
General hauling	Shelling peas
Putting up wood	

Breaking ground, especially new ground, is very arduous work. The task is much harder if the present vegetation is three-year-old alfalfa and the soil is heavy clay. However, the ground can be broken with a round-pointed long-handled shovel. Short-handled shovels are inappropriate because they require too much stooping; they are used primarily for moving relatively loose dirt. Square-pointed shovels will not penetrate with sufficient ease; they are used for making straight edges along a flowerbed. A spading fork is excellent for old ground because it turns the surface and breaks it up simultaneously, but on new ground the resistance is so great that the tines will bend. Old ground can also be broken with a moldboard attachment for hand cultivators.

Fitting the seedbed can be accomplished with ease by steel-toothed rakes. Push cultivators can also be used. The cultivator attachment is beneficial if there is some trash on the surface or if the soil has been rained on and dried out since being broken up. If the soil is heavily impacted, a shovel-tooth attachment for the push cultivator will also prove valuable.

Planting can be readily accomplished by a seeder. The Earthway precision garden seeder and the Lambert seeder, which are of comparable design, have a variety of seed plates which permit planting all our row crops. The current price of these two seeders is approximately $45 and they accomplish all the tasks of planting, covering, and marking the new rows. There is also the one-wheel Plant Right row seeder, which is excellent for planting small amounts of seed. It does not cover or mark the next row.

Potatoes can be planted with a spade potato planter, which sells for approximately $20. For sowing small seed such as alfalfa, clover, and grains, Cyclone makes a drum seeder which also can be used for spreading fertilizer and lime. The drum model comes chest-mounted or wheel-mounted; the chest-mounted seeder is powered by turning a crank, the wheel-mounted by the drive action from the wheels. Cyclone seeders range from $20 to $30. Grass and grain seeds can be sown by a broadcasting horn or by hand.

Cultivating requires a hoe or hand-pushed wheel cultivator. There are also advanced new cultivators, under the trade names of Ro-Ho gardener and the Roe-Boy rotary garden cultivator.

Harvesting requires just a few tools. Most garden vegetables are picked by hand; even dried beans, such as limas, navy, soy, and kidney, can be pulled by hand. Potatoes and onions can be dug by a potato hook, spading fork, or shovel. Lawns can be mowed with a hand-pushed lawn mower. Hay can be cut with a scythe and raked with a hand hayrake. Comfrey can be cut with a pair of pruning shears. Cereal grains can be harvested with a cradle, and corn can be cut with a corn knife.

Processing is a relatively easy matter with a few tools. Cereal grains such as wheat, oats, and barley and dried beans can be threshed with a homemade flail. All that is required is an old shovel handle, a length of chain, and a few eye screws. Corn can be shelled with a hand-cranked Black Beauty corn sheller and walnut huller, which costs about $50, or for smaller amounts by the Decker hand sheller, which sells for about $4.

General hauling can be accomplished with a wheelbarrow, child's wagon, garden cart, or the famous Vermont cart, which is available in three different sizes. Refuse can be moved by a number of different styles of multitined forks and shovels and hauled in any one of the conveyances for general hauling.

Putting up firewood requires only a single- or double-bitted ax. You can also buy a one- or two-man saw to fall trees, a buck saw to saw the wood into appropriate lengths, and a splitting maul to split the wood more proficiently.

Most of these tools can be purchased at your local feed store. Those which cannot be purchased locally can be ordered from the catalogs of the following mail-order stores:

COUNTRYSIDE GENERAL STORE, LTD.
Highway 19 East
Waterloo, WI 53594
1-414-478-2118

CUMBERLAND GENERAL STORE
Route 3 Box 479
Crossville, TN 38555

GLEN-BEL'S COUNTRY STORE
Route 5
Crossville, TN 38555

GARDEN WAY
Charlotte, VT 05445
1-802-425-2121

Often the value of hand tools is contingent upon the user's skill. We refer you to the following excellent articles on the use of the scythe (which also applies to a grain cradle), hayrake, and one- and two-man crosscut saw:

"The Scythe," by Frank Holan. *Country Journal*, July 1979, 43–47.

"Cheap Ways to Rake Hay," by Edna L. Ryneveld. *Countryside*, October 1968, 32–33.

"The Crosscut Saw Manual," by Warren Miller. U.S. Government Printing Office, Washington, D.C.

"The Six-foot, Two-man Crosscut," by Aldren A. Watson. *Country Journal*, March 1980, 53–59.

For an excellent series of articles on hand saws, axes, power saws, and other equipment for putting up cord wood, see *Country Journal*, October 1977, 55–79.

A few words on using other tools: spading new ground is difficult, and any way to do the job more efficiently is generally welcomed. The biggest problem is covering the trash, which is extremely difficult if the trash is four-foot-high pigweed. We suggest that if you do not have a mobile pigpen you make a small mobile poultry pen; the poultry will at least break up the trash well. In lieu of this, spade early in the spring; the trash can then be turned under with a spading shovel or fork. Go in at a slight angle with the point of the shovel away from you. Drive the spade into the soil by placing your stronger foot against the top of the point. Pry the shovel loose by pulling the handle toward you; the higher you pull on the handle, the greater the leverage. Lift up the shovel with the clump of dirt on it and turn it over, placing the topsoil under. The next row is laid against the preceding row. With a little practice, you can also turn under some degree of trash or green manure.

Harvesting potatoes is another difficult task to do by hand. You can use a spading fork by going into the hill at an angle, then lifting the fork full of dirt and potatoes and turning it over on the soil from the direction you are coming. A potato hook is far more efficient. This is essentially a five-tined fork bent at right angles to the handle. Reach across the hilled row of potatoes and pull it toward you. The dirt will pass through the tines of the fork and the potatoes will be moved out into the open. With some practice, this process becomes relatively efficient with a minimum of injury to the potatoes.

Many people do not plant dried beans or grow their own cereal crops because they are not familiar with hand threshing. However, hand threshing is relatively easy with the right tools and a bit of skill. We find threshing beans in the fall to be a very exhilarating experience. The grain and its straw must be thoroughly dry, therefore the grain or beans should not be harvested until well ripened. They should then be cut or pulled and left to dry in the sun. They can then be brought into the barn to be threshed at one time or periodically. However, threshing must be done late on dry days, for the straw will pick up moisture from the air. The necessary tools are a flail, a threshing board, and two large containers; a pitchfork is also valuable. A threshing board made out of particle board or $\frac{3}{4}$-inch plywood is best. It should be about 4-by-6 feet with an 8-inch **V** cut in one end. Nail 2-by-4s around the upper side edges on the three sides which do not have the **V**. Place the threshing board on four cement blocks, one at each corner. Place one of the containers under the **V**, and the fork on the grain or beans.

The flail is made out of a large handle with an eye bolt screwed through the end. Take a short piece of dowl (eighteen inches) about the same diameter as the handle, and screw an eye hook in the end. Attach one end of an eight-inch length of chain to the eye hook on the handle and the other end to the hook on the dowl.

The flail is used by holding the handle and swinging the dowl flatly against the straw on the threshing board. As the piece of dowl comes against the grain, it should be picked up and swung up and over to come down again on the same side with a circular motion on the handle of the flail. Chaff is removed by winnowing—pouring the grain from one container to the other in the open on a windy day. Insects are killed by heating the beans or grain in the oven at the lowest heat for an hour.

All the major kitchen jobs can be accomplished with tools using human power to turn a crank; there are cream separators, butter churns, flour grinders, meat grinders, vegetable choppers, slicers, purers, squeezers and strainers, and pea shellers operated by human power. A very excellent pea sheller can be adapted from a hand-cranked clothes ringer, and you can cut meat with a large meat saw and good butcher knives.

There are a number of miscellaneous tools which are also necessary for homesteaders: a hammer, a crosscut and a rip saw or a power hand saw, a brace and bits or an electric drill, a level, fencing pliers, post-hole diggers, an eight-pound sledge hammer, a steel bar, a block and tackle, a jack (a bumper jack in most cases is sufficient), a length of heavy chain, and a peavey or cant hook (used in logging).

ANIMAL POWER

Oxen should be considered a source of animal power, especially for plowing and hauling. Oxen are castrated bulls, generally dairy bulls; they are powerful and relatively efficient. They are slow, steady workers, ideally suited for breaking ground. Beef bulls can also be used, and it is not necessary to castrate them if they are going to be used on the homestead. This permits the dual function of breeding and draft work.

The use of oxen is gaining popularity, especially in the Northeast. Most of the local fairs have ox-pulling contests. If you are interested, send for an issue of *The Evener*, published in Putney, Vermont. Although this journal is not exclusively devoted to oxen, it is the best journal for oxen enthusiasts. You may also wish to contact county agents in northern Vermont, New Hampshire, and Maine for names and addresses of oxen teamsters.

Horses

In considering animal power, a team of big beautiful draft horses usually comes to mind. For years they have symbolized the American farm. Heavy horses are being rediscovered, and are used exclusively or in conjunction with tractors on many family farms with enough land to support them. However, a team of heavy horses may not be for the homesteader on five acres.

First, consider that a good, well-broken team of mature draft horses costs between $2,000 and $3,000. A new double harness is about $600, and collars and hames an additional $200. And you will need some tools: a plow, $100; a cultivator, $175; a planter, $175; and a mowing machine, $400. With a hitch cart, a ground-driven, small-tractor manure spreader can be used; however, this will cost you about $600. Of course, by doing some scrounging, you may be able to pick up some pretty good used tools. We bought our plow for $15, cultivator for $25, mower for $75, potato digger for $45, wagon for $75, spring-tooth harrow for $10, and disc harrow for $50.

In addition to capital outlay, horses must be fed. Draft horses are heavy eaters; an 1,800-pound horse requires 27 pounds of grain and 18 pounds of hay per day, or 9,855 pounds of grain and 6,570 pounds of hay per year. Feeding a pair of big horses costs about $2,500 a year; if you raise your own feed, it takes two acres of hay. If you tried to grow all your own feed on a five-acre homestead, your team of heavy horses could literally eat you out of house and home.

However, if you are really into horses, your feelings will probably be the determining factor. For horse lovers, there is no sense living in the country if they cannot have a horse around. For the sake of discussion, let us assume that you are a horse lover and the maximum expenditure you can afford is feed for one 1,800-pound draft horse. This permits either one heavy draft horse, a small draft horse, a heavy saddle horse, two light horses, or several ponies, mules, or donkeys. If your primary reason for having a horse is to work it, then it makes sense to get the animal that was developed for work through years of selective breeding, training, and experience: a draft horse. There is nothing stopping you from taking a ride on your draft horse, of course, and you can hitch your draft horse to a wagon, large cart, or sleigh and take Sunday afternoon drives.

A well-broken draft horse can train you to be a teamster. However, if you are not already an experienced horse person, we strongly recommend you get used to a broken draft horse, a saddle horse, or pony before contemplating breaking a colt. You also can get valuable knowledge from a neighbor or from a draft-horse class (we know of one offered at North Adams State College in Massachusetts). As our friend Napoleon says, "A person can get hurt real bad fooling around with a heavy colt." Draft colts generally have good dispositions; however, they are large and you can accidently get hurt without the horse meaning to harm you.

For anyone interested in draft horses, we recommend Maurice Telleen's *The Draft Horse Primer: A Guide to the Care of and Use of Work Horses and Mules*, published by Rodale Press. Telleen also edits an excellent periodical entitled *The Draft Horse Journal*, which can be ordered through the editor at Route 3, Waverly, IA 50677. *The Evener*, published in Putney, Vermont, is also an excellent journal for draft-horse owners.

Harness, collar and hames, rolling stock, and tools are readily available both new and used. For custom-made harnesses, we have had excellent service from:

SMUCKER'S HARNESS SHOP
Box 48
Churchtown, PA 17510

Village Harness Shop
R.F.D. 1
Ronks, PA 17572

For rolling stock, we have had excellent service from:

David L. Fisher
Fisher's Carriage Shop
R.F.D. 3
New Holland, PA 17557

Both new and used harness, rolling stock, and tools can be purchased from:

Phil Stanton
Wild As the Wind Farm
Upton, MA 01568

New tools can be purchased through Glen-Bel's Country Store or Cumberland General Store, both in Crossville, Tennessee.

If your primary reason for having a horse is to ride or drive, then one or two light horses will probably be your preference. We have found that either of our light horses can handle one-horse tools for short periods. We sometimes hitch them double so they can both get exercise. On a five-acre homestead, even with a wood lot, it is hard to provide any horse sufficient exercise by just working them. Actually, even one light horse provides you with a surplus of horsepower.

In all cases, converting a light horse into a multipurpose horse requires adaptation. Light horses are bred and trained to move fast, pulling only a light load, if any. Most breeds adapt equally well, although Arabians and Thoroughbreds seem to be less able to adapt to draft work. There appears to be more of an individual difference within the other breeds. So-called hot or spirited horses do not seem to adjust to draft work. Our big horse is half Arabian and half Appaloosa; our small one, half grade pinto and half quarter horse. They are both excellent multipurpose horses. We broke them to work, drive and ride simultaneously. It is easier to create a multipurpose horse through training than to convert a horse first trained to saddle.

A colt should be started at about a year, assuming it is already halter-broken. Colts vary greatly in the speed with which they can be pushed. Give them sufficient time in each phase before progressing to the next phase. Start by giving them experience with the bit. Put on a work bridle and give them plenty of experience by leading them. Some trainers leave the bridle on for a few days when the horse is in the paddock. When they are used to the bit, get them used to the harness. Then, drive with the long lines a number of times. (Long lines are lines long enough so you won't get hurt if the colt kicks.) Be sure that your colt is ready for the long lines because you are in a very dangerous position. Colts have a tendency to turn and face you when you first drive them; have an assistant handy to help you straighten out the horse. With a few times on the long line, your colt should get the idea.

Now you are ready to hitch to a whiffletree. Have your assistant occasionally pull back on the whiffletree so the colt can get used to the feel of the draft against the collar. After the colt gets used to the occasional light draft, go through the routine of stopping and starting, with your partner putting on considerable draft. As the colt starts to move, have your assistant ease up.

At last, in either a matter of days or weeks, depending on the horse, you are ready to hitch to a drag. This may be a stoneboat (heavy planks fastened together) or a toboggan. However, be sure to pull it on level ground, for it might run up on the horse if pulled down a hill. Stand on the drag and let the colt pull you until he gets used to your weight. You then can increase the weight, but never overload; when he seems to be moving without straining, you are in business. He can do light work around the homestead, but do not tire him.

At this point you can also hitch him to a cart or light buggy. One bit of important advice: do not start him at a trot from a standing position or he will try to move heavy loads at a trot, and then you are in trouble. When he is two to three years old, you can start breaking him to ride. You will find that he will take to a saddle very easily. It is better to break the colt to pull, to drive, and to saddle, in that order.

We have found that a gentle horse with some age can be converted by "negotiation." Older horses may never take to it with the apparent pleasure of a young horse broken to pull, but most are willing if you are patient in teaching them what is expected. Take plenty of time in going through the above steps and be especially patient in applying the draft. Remember, not only is this all new to your older horse, but is also contrary to its natural instincts, training, and experience as a saddle horse. Older horses vary in the ease and speed of adjustment. Many older saddle horses will always have an inclination to bolt and run because this is what they have been trained to do.

We have found that for lack of space and time we have not been able to train our horses to work with enough precision for one person to cultivate or plow. Therefore we have to use someone to lead in doing precise work, even when using one horse. Often it is not possible to find someone who is large enough not to get stepped on. In such a case, we use a stick made from a broom handle with an eye bolt in the end of it. A close snap is fixed to the eye bolt and snapped to the left side of the bit, permitting a small person to lead the horse without danger. Riders can control horses to do precise work; using leaders and riders is a common practice both in this country and in Europe.

If you are going to work your light horse, you will need a well-fitting collar and sturdy harness. Do not improvise; the collar must fit well, even if it means buying a new one at $50.

When it comes to a harness, a little improvisation is possible as long as the end result is sturdy. The basic harness includes bridle, lines, hames, traces, and rump strap. The hames can be cut out of hard wood, the traces made from chain purchased from a local hardware store or lumberyard, and the lines and rump strap made out of sturdy rope. With this harness, you can haul tools such as a plow, spring-tooth harrow, leveler, and cultivator in addition to skidding logs. If you are going to pull a wagon or cart, however, add a britching, holdback straps, saddle, and hold-down straps.

We have found it most satisfactory to use a light driving harness, substituting a work collar, hames, and traces for the driving breastplate. We can go single or double just by changing the lines. We have found that our driving harnesses are sufficient in strength, and keeping up two harnesses instead of four provides a lot more time to enjoy the horses.

Of course, a full range of tools will be necessary. One-horse tools were designed for horses weighing about 1,500 pounds; the heavy horse of a ton or better can pull these tools day in and day out without any strain. A good-sized horse can pull the same tools for most of the work necessary on a five-acre homestead. It takes one of our horses about an hour and a half to cultivate our acre and a half of field corn with a walking cultivator. Plowing is probably the hardest work your horse will do; you won't want to have your light horse pull a one-horse plow over half a day, even with sufficient rests. You can also hitch a team of light horses to one-horse tools to do most jobs, except cultivating. Two horses work well with the one-horse plow, which provides a longer work period with less strain on the horses.

Tools are available new, but restoring old tools provides a great sense of satisfaction. Taking that old cultivator, brushing or sand-blasting the rust off, putting on new handles and painting it, cements a relationship between you, the tool, and the horse that cannot be equaled by purchasing a new tool.

The hitch cart gives you an opportunity to turn your horse or team into a tractor so you can use tractor-drawn equipment. With the cart, you can use horses hauling a light trailer or small tractor manure spreader. A three-gang lawn mower is also about the right size for a one-horse hitch cart. We use the team and hitch cart to pull a one-horse disc harrow. The hitch cart requires wheels, shafts or a pole, and drawbar. Hitch carts may be equipped with a seat.

A pony or two has great potential as a homestead source of power.

These miniature horses were originally developed to work in British mines. They are efficient (easy keepers), powerful for their size, and steady, if broken young. Many are now being bred solely as pulling ponies. The ponies, harnesses, and rolling stock (carts, buggies, wagons, and sleighs) can be readily purchased new or used. The problem is finding tools. We have only seen two tools built specifically for a pony: a very old cultivator which we saw at an Amish farm, and a pair of push lawn-mower heads rigged together as a gang.

We use a pony on our homestead by successfully converting a push-hand cultivator. An enterprising person can easily construct a one- or two-pony hitch cart to which garden tractor implements are attached. It seems very possible to be able to use disc harrows, cultivators (used as harrows), gang lawn mowers, tip carts, and yard sweepers. A very enterprising person might even be able to convert a moldboard plow with a drawbar hitch.

Pigs

One of the most difficult tasks on the homestead is breaking new ground, and there is no better animal for this purpose than a pair of hogs in an eight-by-ten mobile pen. They simultaneously open the soil, eat the weeds, and fertilize the ground. They can even be encouraged to root out small saplings by throwing shelled corn around the roots. As they dig for the corn, they will root out the sapling, occasionally pitching the sapling right out of the pen.

TRACTORS

Unless you are very attracted to horses or ponies, you should consider other sources of power. If you are not knowledgeable about horses, a great deal of time and energy will be required to become a teamster. In addition, horses are always present; they always demand your attention. A tractor can be shut off and forgotten.

If you decide on a tractor, take into consideration the amount of work you are willing to do with muscle power, and how much money is available for capital investments. Some homesteaders are interested in using various configurations of the three types of power: muscle, horse, and tractor.

Deciding whether to purchase a new or used tractor is based on mechanical interest, mechanical skill, and working capital. If you are mechanically minded and gifted, classes in small-engine repair and auto mechanics will teach you to repair and maintain a used tractor.

Unlike cars, tractors are made to last, as witnessed by the great number of 1938–1947 Ford 9Ns still in operation. The two-wheeled tractors and smaller air-cooled four-wheel tractors do not last as long. Essentially, the engine goes before other parts. For a few hundred dollars, the old engine can be replaced, and the need to buy a new tractor eliminated.

We saw advertised a sixteen-horsepower two-wheel tractor with mower, snowblower, fertilizer spreader, trailer, and tree sprayer attachments for $2,200; another with a disc harrow and plow with a good motor for $250; and still another with snowplow, land plow, and cultivator with a five-horsepower engine for $200. Two-wheel machines are available with diesel engine and attachments such as a rear-end tiller, sickle-bar mower, hiller, land plow, grader blade, log splitter, snowblower, grinder, tip cart, and grain binder.

There are a number of other two-wheel power tools. Of greatest popularity is the rear-end rototiller. Its primary purpose is breaking ground and fitting, but it can be used very effectively for cultivating and even hilling potatoes or corn. Add a mower and snowblower combination and you have the power for five major homestead tasks: breaking ground, fitting, cultivating, mowing hay and grain, and removing snow.

Two-wheel tractors seem to be losing out in popularity to the small four-wheel riding garden tractors. It is a little perplexing for the homesteader to see the suburbanite mow his lawn with a four-wheel garden tractor and then spend half an hour jogging around the neighborhood for exercise.

Two-wheel tractors are fine for up to two acres, but for five acres you may feel something more powerful is justified. Consider a small used four-wheel tractor; if you do not demand mounted equipment as opposed to a pull-behind, then the range in used tractors is great. Another possibility is a jeep or other four-wheel-drive road vehicle. Some of the oldest tractors are still available. It is not unusual to purchase an old workhorse in pretty good shape with a trailer plow and cultivator for as low as $700. You can also adapt horse-drawn equipment to a draw-bar tractor; for instance, if you have a horse-drawn plow you can chain the plow to the axle of the tractor; one person drives and another plows. Two people can cultivate in a similar manner. Adapting most horse-drawn equipment such as the plow, cultivator, and mowing machine requires two people, one to operate the tool and the other to operate the tractor.

Three advantages to using mounted equipment are maneuverability, gas economy, and ease of transport. Mounted plows, cultivators, and mowing machines were first added to smaller tractors; the equipment was raised and lowered by a lever. These tractors were later equipped with hydraulic systems and older models were converted by hydraulic kits. A greater range of tools was added, with fingertip raising and lowering; for instance, a plow, two- and four-row cultivators, two- and four-row corn planters, a seven-foot sicklebar mower, bean puller, snowplow, and manure loader, as well as a standard power take-off and pulley to operate a saw rig. One of these used tractors with mounted land plow, snow plow, cultivators, and mower in good condition sells for $2,200 to $3,500.

The top of the line for homestead power is a three-point-hitch tractor. The classic three-point-hitch tractors are the Ford and Ferguson. The first of them were manufactured by Ford in 1939 as the 9N model and distributed by Ferguson Companies. Ferguson had been in partnership with Henry Ford manufacturing earlier tractors. In 1947, significant improvements modified the model and it came out under the name of 8N; this was produced until 1953, when it was replaced by the Golden Jubilee model NAA. In 1948, Ferguson started producing a model very similar in appearance and engineering to the 8N under the model code TO-20, which was produced until 1951, when it was replaced by the TO-30, which was manufactured until 1954. The patent finally expired and many of the other tractor companies picked up the three-point hitch. This now permits interchangeable equipment and three-point-hitch tractors. Parts and owner's manuals are still readily available even for the Ford 9N.

The range of mounted equipment is limitless, from the standard two-bottom plow through front-end loaders with down-pressure buckets, rear-mounted hoists and cord-wood saw and splitter combinations. The added advantage is that these are such good tractors and so highly valued by their owners that there is an abundance of quality units around. In the Northeast, a good unit sells for up to $3,000 without equipment. This price compares very favorably with the $5,000 to $10,000 for new tractors.

The homesteader's first energy consideration is human power. All major homestead tasks can be performed by people power, if there is enough of it. Oxen or horse lovers who need additional sources of power will probably choose animal power: oxen (and beef bulls), heavy horses and mules, light horses, or ponies and donkeys.

Homesteaders not into draft animals will probably consider a tractor their primary source of supplemental power. Those who are mechanically inclined will choose used tractors.

A wide range of tools is available for all power sources. Homesteaders can establish the most appropriate system of energy and tools for their particular operation and effectively modify the system as conditions and circumstances change.

N

10

POSTAGE–STAMP HOMESTEAD

VISITORS from the city or suburbs often say after visiting us, "You've got a nice place here. We would like to homestead, but we just don't have enough room in town." We also know a few people who did not let the lack of space stand in their way; one couple raises four feeder pigs annually on a small lot in the center of town. These friends got us thinking about what we could have raised on our half-acre site in town. After extensive research and deliberation, we found we could have done quite well on a postage-stamp homestead of half an acre or less.

This chapter is for those who live on a small lot, but are interested in a family hobby that provides a creative and healthful outlet, saves money, and contributes to the good life. A postage-stamp homestead may be an end in itself, or a training camp to learn homesteading in order to prepare for larger-scale homesteading or farming.

MORE FROM LESS: SYNERGY AND SCALE

With Dick's farming experience in large commercial cattle operations, it has always been hard for him to think in terms of small-unit production. In extensive dialogue with other people, especially novices to homesteading, we have found that many people have the misunderstanding that large areas are needed to produce food. Buckminster Fuller, a leader in curtailing this tendency to think on too grandiose a scale, urges "more from less." His words tell us that if we live on a half-acre homesite, we should see what we can do right on our half acre and to rely on all our strengths and resourcefulness.

One of the secrets of getting more from less is synergy, another of Fuller's concepts. Synergy is the cooperative working together of different units, with a result greater than the sum of the parts. This is clear when one examines livestock, fruits, and vegetables; the droppings from rabbits enhances the growth of fruit and vegetables through composting. Synergy makes postage-stamp homesteading possible; a well-balanced ecological system in which maximum production and satisfaction are achieved through thoughtful interrelation of the various components.

There is a tendency for homesteaders to overextend because of the success of a project and what we call the Bambi syndrome. If one is successful with rabbits and finds it difficult to slaughter the "cute little bunnies" (the Bambi syndrome), he or she may well become inundated with rabbits. This can happen with any homestead component, including chickens, ducks, goats, bees, vegetables, and fruits. A five-acre homestead may provide sufficient slippage to allow a venture's going out of scale without disaster, but on a postage-stamp homestead, all components must be integrated precisely. Determine the appropriate scale, then stay with it. On the postage-stamp homestead, scale is determined by that scarce commodity, space. It is essential to allocate space for everything from livestock to small hand tools. The thoughtful use of limited space is not only efficient, it is also esthetic.

People consider esthetics when planning their homesites, and the result is frequently a pleasing arrangement of trees, shrubbery, and flowers; pretty, but of no or minimal food value (day lilies are delicious). On a postage-stamp homestead, it is important to consider the esthetics of utility. There is a beauty in a peach tree; it is beautiful blossoming and beautiful when the fruit emerges. And there is an additional beauty in eating the fresh peaches and preparing them to enjoy all winter long. Similarily, there is beauty in a honeysuckle hedge, but there is an added beauty in a fence constructed of grapes. Three softly cooing white Muscovy ducks leisurely cropping quack grass from around your tomato plants is another example of the esthetics of usefulness on a postage-stamp homestead.

HOMESTEADING IN THE CITY

The biggest constraint on postage-stamp homesteading is not space: it is laws and ordinances. Go to your city or town clerk's office and examine the ordinances relating to homesteading. Of primary concern are those pertaining to livestock, but also examine any ordinances relating to ponds, bees, fencing, and planting trees and vegetable gardens. The ordinances may prohibit any of these on your homesite, or they may be merely regulatory, specifying how the item is used. A prohibitory ordinance specifies that no domestic farm animal is permitted on any homesite within the boundaries of the jurisdiction; a regulatory ordinance states that animals may be kept if they are not offensive to neighbors through noise or smell. Ordinances vary from jurisdiction to jurisdiction, so determine exactly what ordinances apply to you. It is valuable to make and keep a copy of them.

Many ordinances that regulate or prohibit livestock in urban areas were passed years ago. Now, people are more sophisticated about animals; many people know which animals are appropriate for small space and dense areas, and how to care for them so they will not be offensive. However, it is doubtful that ordinances will be readily changed to accommodate postage-stamp homesteaders, and it is also unlikely that government representatives will be knowledgeable about the subject. Written ordinances tend to be conservative; a law prohibiting riding a mule faster than five miles an hour down Main Street is presently in effect in one very progressive urban area.

What should the prospective homesteader do about a prohibitive ordinance? There are three possibilities: seek a variance in the ordinance (an exception made for you); wage a neighborhood campaign for repeal or alteration; or go ahead and be prepared to deal with the consequences. The decision can only come after a thorough investigation of the ordinances in your jurisdiction.

Whatever alternative you take regarding prohibitive or regulatory ordinances, the essential issue is the same: having your postage-stamp homestead accepted by your neighbors. A good-neighbor plan is necessary. Become well acquainted with any neighbor who might be affected by your homestead either through sight, smell, or sound. Gain the reputation of being considerate and helpful; offer your labor and services. Add units to your homestead only as fast as you and your neighbor can assimilate them. If you start with a vegetable garden, do not move on to something else while the garden becomes overgrown with weeds, and the produce cannot be effectively processed and used. Share the planning, as well as the produce, with your neighbors.

Let us assume that you have planted fruit trees and wish to add a vegetable garden and livestock. You are going to start with rabbits and the appropriate vegetables—lettuce, tomatoes, and radishes. You could invite your neighbors over for a rabbit cookout, and as your guests partake of the delicious repast, talk about the value of rabbits, both as food and for helping vegetables and roses grow. Get as much assistance and advice as possible. After a week or so, make the rounds and mention to your neighbors that you have decided to raise rabbits.

Nothing succeeds like success. Sharing the produce grown with rabbit manure and a few rabbit cookouts or dinner parties will solidify your neighbors' support.

PLANNING

What do you really want to accomplish on a postage-stamp homestead? A variety of objectives exist: perhaps you wish to grow abundant high-quality vegetables and fruits free from pesticides; or you may wish to raise chickens and goats for fresh eggs, milk, and cheese, as well as to provide necessary organic fertilizer for fruit trees and vegetables. You may wish to raise Bantam chickens to help control insects and for the fun of watching them.

On the other hand, if you are an animal lover and your primary objective is to have animals, you may wish to raise rabbits or chickens or Muscovy ducks, and you may wish to sell the produce—meat and eggs—either locally or to wholesalers.

Once you have determined your objectives, evaluate your site's strengths and liabilities. Each site presents its own set of design problems; consider the following:

- Shape
- Terrain
- Location of trees and boulders
- Location of the house
- Location of rooms and exits
- Location of existing shelter for animals (garage, barn)

Lay out the site and buildings on graph paper drawn to scale. Make a dozen copies to permit experimenting with a number of arrangements. Even though the implementation of new components should be incremental, an overall master plan is essential.

Next, make an inventory of all your resources, including money, space (both covered and uncovered), time, and labor. We suggest that this be done in writing, with realistic quantities assigned to each item.

We have friends who live in a $200,000 mansion on a five-acre site. The couple are avid homesteaders. They want a sense of independence by producing their own quality food free from pesticides. They also enjoy the physical work of maintaining the homestead. Money is no object to them; for the rest of us however, cost-benefit analysis is important and necessary.

Keep accurate records and evaluate each project annually; the components you select, as well as the way you maintain them, are dependent on cost considerations. For instance, in comparing three Muscovy ducks with four rabbits, the following must be considered: four rabbits from good stock cost $40, and a three-level rabbit cage with six compartments, feeders, and waterers costs about $180, a total of $220. If this seems steep, but you are interested in rabbits and can afford $40, you could make suitable hutches out of wood, used screening, and wire for $5. If you are really strapped for money, you might prefer ducks. Three young Muscovy ducks cost $6, and allow $1 for wood and nails to build nesting boxes. In terms of benefits, three rabbits produce about 300 pounds of meat per year for the freezer; two duck hens will produce about 240 pounds. It also costs about three times as much to feed the rabbits.

FRUIT

Let us first consider fruit trees; these include cherry, pear, plum, peach, nectarine, and apple. Always plant at least two of a variety for cross-pollenation. Fruit also includes berries—raspberries, blackberries, blueberries, and elderberries—and do not forget grapes—red or Catawba, green such as Niagara, and Concord. Fruit provides valuable nutrition and can be eaten fresh, canned, or frozen, used for fruit leathers, and in wines, liquers, and cordials.

Fruits are a very efficient use of land. Properly situated, they use very little land and can enhance the site by providing attractive borders. Fruit residue also enriches the compost pile, which in turn aids fruit production.

Fruits are excellent along perimeter fencing; raspberries and wire fencing go particularly well together, and grapes can also be used for barrier fencing. Trees can be planted along walls and solid fences. The espalier process (forcing trees to grow against a wall) is very attractive and conserves ground space. Plant the tree as close to the wall or fence as possible. As the branches grow against the wall, prune them to form designs against the wall. This is especially beneficial if you have walls or solid fences on the north side of your site so as not to shade your vegetables.

Equipment

Start with canning gear and paring knives; you may also want a cherry pitter, apple peeler, and perhaps a cider press (if you would like to make cordials and wines, the press is invaluable). You will then need a fermenting bucket and five-gallon fermenting bottles.

A dehydrator, either homemade or commercial, is essential for processing fruit leathers. Fruit picked in the summer can be made into leather that provide nourishing and delicious snacks all year round. Dehydrators can be used in conjunction with stoves, separate heating units, or the sun.

In addition, you may want a two- or five-gallon sprayer for insecticides. A ladder, pruning shears, and pruning saw are also needed, as are a wheelbarrow or yard cart, shovel, and hoe. Get a five-gallon pail or a hose to water your fruit trees when they are new or during droughts.

THE POND

The homestead pond is invaluable. It can produce more food per square foot than the same amount of space used for animal production. Fish provide a valuable supply of protein as well as variety to the diet, and offer an excellent leisure-time activity. The pond is a great advantage for ducks and other waterfowl, and enhances the beauty of any site.

The homestead pond should be at least twenty feet in diameter; an ideal pond is elliptical, about twenty by thirty feet, but different shapes can be used. Make it at least six feet deep at its deepest end. Stock the pond with pan fish, such as blue gills and bass, along with a bottom-feeding fish such as carp or catfish. The pond provides two great challenges: keeping it fished out, so it doesn't become overpopulated; and keeping an area free of ice in the winter to provide sufficient oxygen for the fish, by purchasing a wind-operated water agitator, or periodically breaking holes in the ice. The handy person can make his or her own agitating device. The fish will occasionally need to be fed a small quantity of grain. Lastly, it is important that the pond be securely fenced. This is no problem if the pond is in the backyard. Homesteaders considering a pond should contact their county extension agent for advice.

MUSHROOMS

Mushrooms are valuable for the postage-stamp homestead because they require a minimum of labor, produce an abundance of nutrition, need very little space, and are valuable users of compost. Mushrooms can be eaten raw, mixed with salads and dips, cooked in casseroles and soups, stuffed, and sauteed.

Mushrooms can be grown in boxes about one foot deep, three feet wide, and four feet long. Store them in dark, damp areas, where the temperature runs around 60°F (16°C)—in the basement, under the stairs, or in corners of outbuildings. Fill the boxes with a composting combination of hot manure (preferably horse manure) and straw. Avoid woodchips or sawdust. Once the spores are established, the boxes will produce crops of mushrooms for about six months. An excellent guide is Jo Mueller's *Growing Your Own Mushrooms: Cultivating, Cooking, and Preserving,* Garden Way Publishing Company, Dept. A368, Charlotte, Vermont.

THE GARDEN

The garden can provide almost 100 percent of the family's diet. It is also efficient land use, which with proper methods can be made to produce abundantly. In China, it is expected that one acre of ground, in addition to livestock and pond produce, can totally sustain a family of four. The garden can provide fresh vegetables almost all year round; with a greenhouse, you can literally grow fresh vegetables every day of the year. In the Northern states, begin your vegetable garden in April; grow broccoli and cabbage (from transplants), endive, kohlrabi, lettuce, onions (from sets), parsley, peas, radishes, spinach, turnips, and early potatoes. During May, add carrots, cauliflower, beets, onions (from seeds), parsnips, and Swiss chard. A few weeks later beans, corn, late potatoes, and tomatoes can be planted. In the early part of the summer, add lima beans, canteloupe, celery, cucumbers, eggplants, pumpkins, pepper plants, winter potatoes, squash, tomato plants, and watermelons. For a second crop in the fall, seed the garden in late June and early July with beets, broccoli, cabbage, cauliflower, kohlrabi, lettuce, radishes, spinach, and turnips. Small areas throughout the homestead can be used for herbs. The garden can also be used to produce livestock feed. Ducks will be happy to use extraneous grasses as food. Also, a very little area can produce an abundance of comfrey, which is used to supplement chicken and goat feed.

The garden goes exceedingly well with livestock, forming an ideal ecological system. Rabbits, for example, provide the best manure available and earthworm castings (excreta).

By using the French intensive method of gardening, a small space can produce three or four times the quantity of vegetables by conventional methods, especially with the support of good compost. Fences and walls can be used profitably for growing peas, pole beans, cucumbers, and squash. Strawberry pyramids can add to space efficiency.

The French intensive method is relatively simple. It requires much human muscle, but it is healthful exercise. It has been estimated that a person can produce $10,000 annually from produce of one-tenth of an acre.

Beds are approximately five feet wide, with narrow paths in between. Mark each bed with string. Dig a foot-deep trench along one end of each bed and put the removed soil in a cart or wheelbarrow. Loosen the earth in the trench with a spading fork. Continue throughout the bed by moving the topsoil over to the area where the soil has just been removed. When the bed has been completed, there will be a trench at the end. Empty the wheelbarrow full of soil you removed when you began the bed into the last trench. Then cover the bed with a layer of compost and shape and smooth it. If you wish, you can raise the beds with landscape timbers or creosote-treated boards. You are now ready to plant. Rotate the beds so that heavy nitrogen users (corn and beets) are preceded by legumes (peas and beans).

A greenhouse should be seriously considered. In addition to providing year-round vegetables, greenhouses provide excellent areas for growing seedlings and trap solar heat for part of the house. Greenhouses are of varying sizes and shapes. Construct it on the south side of the house to maximize sunlight. Different transparent coverings can be used, including Plexiglas, glass, and polyethylene. It can be as permanent or as temporary as you wish. Rock floors and water-filled oil drums painted black add to its passive solar-heating capacity. The greenhouse also requires a screen door and windows to provide sufficient ventilation during hot periods.

Be sure to intersperse a few flowers in your garden, especially marigolds, for protection from insects. Day lilies and Jerusalem artichokes are both attractive flowers, and a delicious source of food. Be sure to keep weeds under control for efficient space use and pleasing appearance.

If you decide not to use a greenhouse, you will need a cold frame, which can be produced very inexpensively by using old windows and scrap lumber.

You will also need composting bins. Their size depends on the amount of refuse you possess. They can be constructed out of boards, cement blocks, small-diameter posts, or welded wire, such as standard two-by-four mesh. Ideally, the composting bins should be solid to two feet up from the ground. This permits red worms to assist in the composting process without their taking off for parts unknown. Bins should be under cover; use black polyethylene to cover the surface when not in use. Place them close to the major livestock areas, especially the goats. A mixture of manure, household garbage (except for fats and meats), and plant debris is used to build up the compost system. With this combination, the odor is not too bad. Old established compost heaps can be used to grow pumpkins and melons. A compost barrel or garbage can can be kept in the basement of the house. Make some holes in the bottom of the container, put in a little peat-moss, soil and a little garbage, and dump in your order of red worms—about 1,000. Add kitchen waste and occasionally a little rabbit manure and soil. Do not mix these items together; let the worms do the work.

Storage equipment will also be necessary. A large deep freezer is ideal. Years ago we bought a used eighteen-cubic-foot freezer chest for $50. It is still going strong and looks like it will continue for the next ten years. You will also need sand containers for root crops such as turnips, carrots, Jerusalem artichokes, and rutabagas; they can be boxes or plastic or galvanized garbage cans. We like to keep one container separate and unused. At the end of the season we slowly dump the sand into the empty container, pulling out any left-over vegetables. Later we gradually fill the emptied container with layers of vegetables and sand.

Few tools are needed for a greenhouse and intensive gardening. The basic ones are a shovel or spade, a spading fork (which can also be used to dig potatoes), a hand wheel cultivator, a triangle hoe, and a trowel. Plastic or wooden flats are needed in which to grow seedlings. Either a cart or a wheelbarrow is necessary; garden carts (lightweight bicycle-wheel carts) are nice but expensive, and a little wide for French intensive gardening. A wheelbarrow is cheaper and takes up less space.

If you plan to freeze and can foods, you will need a blancher and canner; a juice extractor is also useful. There are a number of hand and motor-powered types of each appliance on the market.

HONEY BEES

Honey is used as a sweetener and it has more food value than refined sugar. Honey bees pose no problem for livestock or garden; dead bees can be used to feed chickens and ducks as well as pond fish. Bees need very little space. However, they must be situated so that their flyway does not conflict with people or animals. It is also valuable to provide a somewhat concealed area. The flyway must be open to the south. Two hives are the maximum. Before deciding on bees, find out if any of your close neighbors are highly allergic to bee venom. In addition to the hives, you need a smoker, gloves, hat, and hive tools. An extractor is ideal, but somewhat expensive; you can make your own extractor or join with other beekeepers in sharing one.

RABBITS

Rabbits are one of the most efficient and enjoyable animals to raise. Their meat is delicious and can be prepared many ways, and their droppings provide both the richest and safest manure available and the best environment for raising red worms. Rabbits do well among fruit and vegetables, and do not conflict with other livestock—except chickens.

For equipment and space needed, see Chapter 3.

GOATS

Goats are extremely efficient animals and ideal to raise in small areas. One goat takes up less room than a medium-sized dog, is lovable, and keeps a family of five in milk. The delicious cheese can be made for 5¢ a pound and goes exceedingly well with homemade wine. For equipment and space required, see Chapter 4.

YARD BIRDS

Yard birds are valuable as food and help to control insects. Yard birds include ducks and chickens that are permitted to run loose. With fences and gates secured, yard birds are a labor-free addition to the postage-stamp homestead. We recommend two or three Muscovy hens and a drake; Muscovies are ideal because they are grass eaters and relatively quiet. Their hissing is inoffensive and their cooing is very pleasant. Although Muscovies swim, they are not overly fond of water. Their use of the pond will be sufficient to help fertilize it without polluting it.

A trio of attractive ornamental Bantams would also be nice. You may have to forego the rooster if your neighbors are close and complain, but the hens will provide excellent eggs which are exceedingly delicious hardboiled or pickled.

For equipment and space requirements, see Chapter 5 and 9.

LAYERS AND BROILERS

A flock of eight to ten layers can provide enough eggs for a family on a year-round basis, as well as providing some very good stewing chickens every year or two. A unit of twelve layers will provide enough eggs to sell, and thus pay for the feed.

If you add a rooster, you can also use eggs to grow your own broilers. It may cost as much to raise broilers as to buy them at the market, but your homegrown broilers will be far superior to the force-fed, commercially raised ones.

For equipment and space required, see Chapter 4.

MISCELLANEOUS EQUIPMENT AND AIDS

In addition to the equipment mentioned above, a station wagon or an enclosed pickup or trailer is necessary. You need an enclosed conveyance to transport your goat to a buck and to pick up your feed and supplies. A small trailer on which you build a removable cover is sufficient for these projects.

An additional concern is to locate a nearby feed store. You may be able to purchase your feed from a local farmer; this saves you gas and travel time.

It is also valuable to subscribe to homesteading periodicals. One of the most valuable for the postage-stamp homesteader is *Countryside.* In some areas there are homesteading organizations. Keep your ear to the ground; perhaps there is one in your area. Do not forget to visit your county agent.

Most postage-stamp homesteaders can accommodate the following components:

- **Fruit**
  - Trees: Apple, pear, peach, plum, cherry
  - Bushes: Grape, raspberry, elderberry, blueberry, strawberry
- **Vegetables**
  - Greenhouse grown
  - Open space (French intensive)
- **Pond**
  - Carp, catfish, bass
- **Mushrooms**
- **Bees**
- **Livestock**
  - Rabbits, Goats
  - Yard birds (ducks and Bantams)
  - Layers and broilers

The necessary equipment is minimal and rather inexpensive. A closed convenience is the most expensive, but perhaps you already have a station wagon. A little investment of time, planning, effort, and money provides returns in satisfaction, healthful exercise, food, and savings.

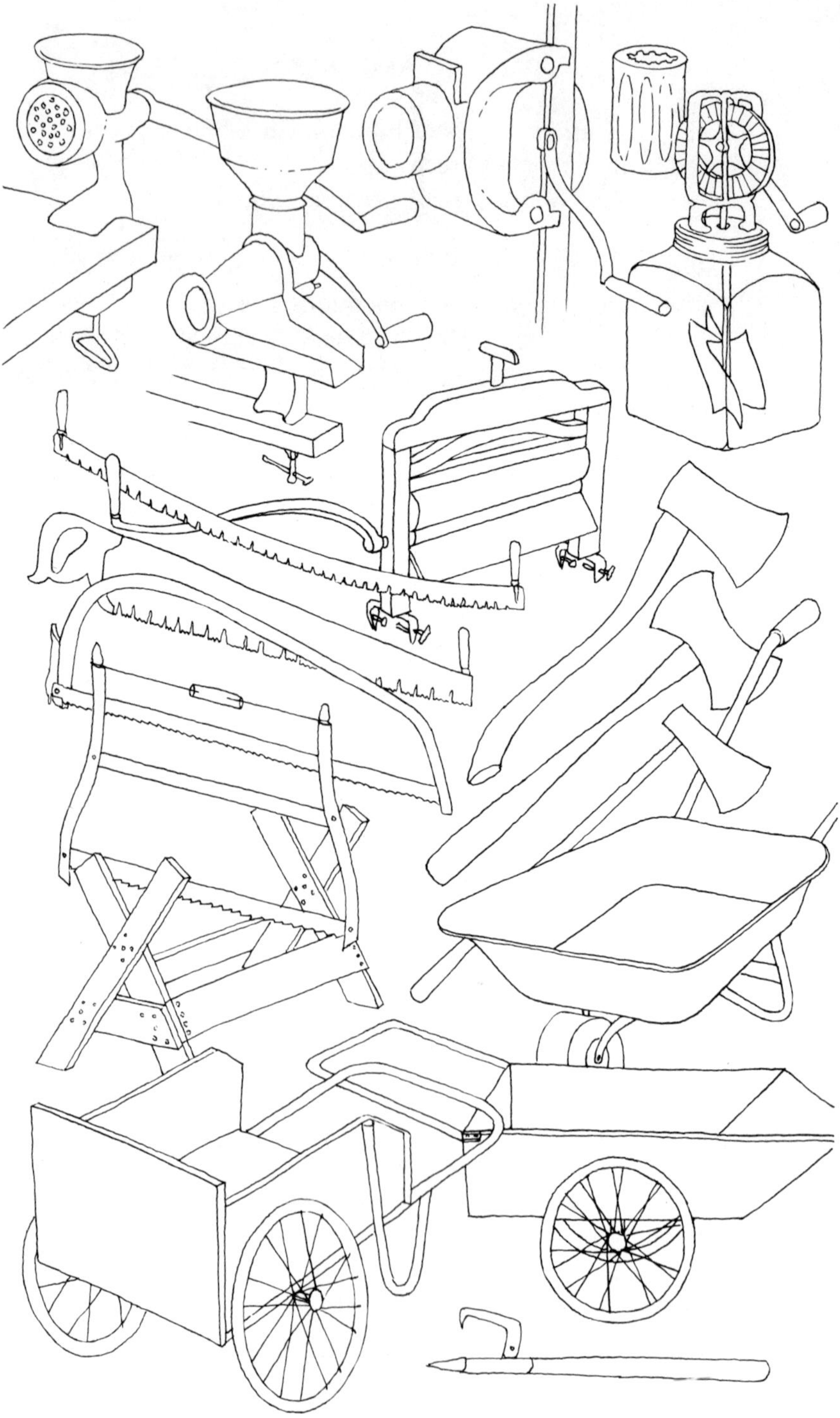

11
HOMESTEADING REFERENCE GUIDE

THIS CHAPTER is a handy alphabetical reference source for those facts homesteaders want in a hurry. We have also included some important subjects that did not seem to fit anywhere else in the book, but which should be part of a homesteader's knowledge.

BEDDING MATERIAL

There are a variety of materials that can be used for bedding. The absorptive quality of some of the more common materials is provided in the following table:

Bedding Material	Pounds of Water Absorbed per Hundred-weight of Dry Bedding
Peat moss	1,000
Oat straw, chopped	375
Vermiculite	350
Wood chips, pine	300
Wheat straw, chopped	295
Oat straw, long	280
Peanut hulls	250
Sawdust, pine	250
Wheat straw, long	220
Sugarcane bagasse	220
Woodchips, hardwood	150

BREEDING

The variety of animals found on a homestead can make it confusing to remember the breeding schedules for various animals. To assist you, we have provided the following table:

Animal	Age at Puberty	Heat Cycle (Days)	Average Length of Heat
Cat	6–15 months	15–21	9–10 days
Cow	4–18 months; first bred about 15 months	18–21	18 hours
Dog	6–18 months	—	21 days
Goat	7–8 months	19–21	2–3 days
Horse	1 year	Variable; about 22	4–12 days
Pig	5–8 months	20–22	2–3 days
Rabbit	5–9 months	—	—
Sheep	7–8 months	14–19	30–36 hours

BUTCHERING

Butchering is among the least desirable jobs on the homestead; you will get more volunteers to clean out the barn than to help butcher. The steps for butchering rabbit and fowl have been addressed in their respective chapters, but what about the larger animals? You may truck your animals to the nearest slaughterhouse, which is what most homesteaders do, or you may do the butchering yourself. A home abattoir requires a large barrel for hot water, pipe or hammer, appropriate knives, pulleys and ropes, scrapers, and a good meat saw.

Two other necessary items for butchering are time and skill. The skill can only be acquired by working with somebody who already is an expert. We do not recommend that you attempt to butcher or cut up meat on your own without having learned the skills first-hand. Much can go wrong; improper killing can result in cruel and painful treatment of the animals, and improper cutting can cost you in unusable meat.

You need to be familiar with meat cuts to properly order what you want from a slaughterhouse. Most good cookbooks have charts at the beginning of the meat section showing the various retail cuts.

CHEESE

Cheese production is an excellent way to use up surplus milk. Cheese making requires milk, rennet tables, inexpensive equipment, time, and patience. Cheese making is an art, and each cheese wheel you make will taste different. If one wheel is not to your liking, try again and perhaps refine your procedures or vary your recipe.

Cottage cheese is the easiest cheese to make. Put one gallon of milk in an enamel kettle or pail (aluminum is affected by the acid in the curd) and place the milk pail in a larger kettle containing water, making a double boiler. Warm the milk to 86°F (30°C), remove from heat and add the rennet according to the directions that come with it. Stir with a wooden spoon for a minute. Let the milk stand until a curd forms firm enough to be lifted by a finger, about thirty to sixty minutes. Cut the curd with a knife long enough to reach the bottom of the pail. Slice across the whole curd in strips straight down about $\frac{1}{2}$ inch apart, then cut $\frac{1}{2}$-inch intervals at a right angle. Gently stir the curd and cut any pieces you might have missed.

Let the curd rest for fifteen minutes, then place the curds back into the larger kettle and very slowly reheat the cheese to 100°F (38°C); this should take about forty minutes. Frequently stir the cheese gently to keep the curds from lumping together. Once it reaches 100°F (38°C), remove both pans from the heat. As the curd cools, keep stirring often to prevent the curds from sticking together. The curds contract as they cool and this forces out more whey. Do not let the curds cool to below 86°F (30°C). The curd has reached the desired firmness when it breaks apart easily, showing little tendency to stick together when a small amount is squeezed in your hand and released quickly.

Let the curds rest a few minutes and then pour off much of the whey. Drain the rest of the whey by using a strainer lined with cheesecloth. Stir the curds to keep them from lumping together. The curds are ready when they have a rubbery texture and squeak when you chew a piece. Rinse the curds in cold water to remove the remaining whey; you may then add salt to taste, and cream or milk, if you wish. The amount of cream or milk to use depends on how creamy you want the cottage cheese.

Save the whey for the pigs or chickens, for whey lemonade (one quart whey, six tablespoons sugar or honey, and juice of two lemons), or for a Ricotta-type cheese.

To make hard cheese you will need a cheese hoop or form, which can be made from a two-pound coffee can. Remove the top completely and punch holes in the bottom with a nail, punching from the inside out to ensure a smooth surface next to the cheese.

You also need a follower, a round piece of wood about an inch thick. Use the lid you cut off for a pattern and make the follower so it fits loosely enough to slide up and down in the can. The function of the follower is to force the wet curds together, squeezing out whey and forming a solid wheel of cheese.

A cheese press is needed to exert pressure on the follower. You may buy or make a cheese press or use weights or bricks. Generally speaking, the longer a cheese is pressed and the heavier the weights, the harder the cheese.

Hard cheeses are made by following the same steps used to make cottage cheese. After draining the whey in a strainer, add about

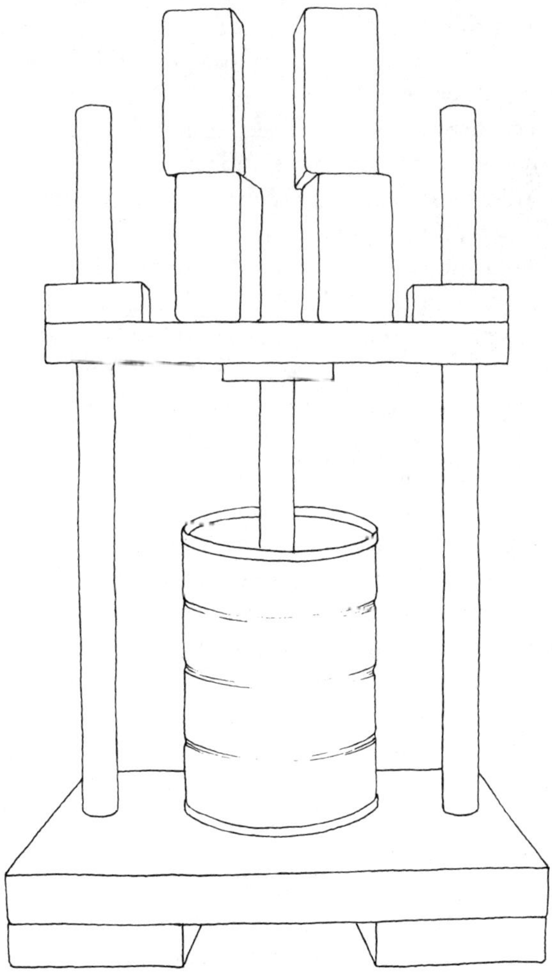

Cheese press

three teaspoons of coarse salt per pound. Once the curd has been lowered to a temperature of 86°F (30°C), put the cheese in the cheesecloth-lined cheese hoop or form; make sure there are no wrinkles in the cloth along the sides of the form. Press the cheese down and out to the sides with your wooden spoon. The cheesecloth will stay in place more easily if you use a rubber band to hold it around the outside of the form. The cheesecloth must cover the bottom as well as the top of your cheese. Set your form in a pan to catch the whey that will be squeezed out and place the follower on top of the cheesecloth-covered cheese.

You are now ready to press. The amount of weight varies with the amount of whey left in the curds and the amount of cheese you are pressing. Allow approximately ten pounds of pressure per pound of cheese. After an hour or two, remove the weights and follower and lift the cheese and cheesecloth out of the form. Rinse off any fat on the surface and wipe dry. Trim off any excess amount of cheesecloth, so that you can surround your cheese without much overlap. Place the cheesecloth-wrapped cheese back in the form, replace the follower and press with the weights for about twenty-four hours at an ideal temperature of 60°F (15°C).

After twenty-four hours, remove the cheese from the press and form. Take off the cheesecloth and wipe the cheese with a clean, dry cloth. Wash the cheese in hot water to help develop a firm rind. Put the cheese on a shelf in a cool, dry place, about 55°F (13°C). Wipe the cheese and turn it over daily. In about five days the cheese should have formed a good rind. Continue to turn and dry cheese daily until eaten, unless you cover it in paraffin after a rind forms.

To cover the cheese with paraffin, melt the paraffin at 210°F (98°C) in a double boiler. Dip half the cheese in the paraffin for about ten seconds, then remove it and let it cool for a couple minutes. Repeat the dipping procedure with the other half of the cheese, making sure you have left no undipped areas. Store the cheese in a cool place for about six weeks. If you want a sharp flavor, store the cheese for three to six months.

Yogurt is also a good use for surplus milk. An easy yogurt recipe is to mix $\frac{1}{3}$ cup powdered skim milk into 1 quart of fresh milk and heat to 112°F (45°C). Remove from the heat and mix $\frac{1}{2}$ cup warm milk into $\frac{1}{4}$ cup prepared yogurt (used as a starter) that has been warmed to room temperature. Stir the diluted yogurt into the milk and then pour into a jar or covered bowl.

Wrap the mixture in a towel or blanket and set in a warm place overnight or for a minimum of six hours. You may also set the yogurt in your oven, along with (but not in) a pan of hot water to raise the temperature. Store the finished product in the refrigerator. Remember to save $\frac{1}{4}$ cup of your yogurt to start your next batch.

COMPOSTING

Composting is a hallmark of homesteading; it epitomizes the adage "waste not, want not." Compost is decayed organic matter; composting is the restructuring of the organic matter for maximum value as a fertilizer. Compost adds important nutrients to the soil—nitrogen, phosphorus, potash, and trace minerals. It also neutralizes the soil; overacid soils become more alkaline and overalkaline soils become more acid. Clay soils become more crumbly and sandy soil finer, which improves the soil's water-holding potential. Each compost granule works like a sponge, holding almost its own weight in water, releasing the moisture as it is needed. For similar reasons, compost protects soil from wind and water erosion. Lastly, it aids in providing oxygen and necessary bacteria and fungi to the soil.

Nature is the foremost composter. It is always at work changing the structure of organic matter. Unless dead organic matter is removed from the homestead or burned, nature will turn it into compost. However, the homesteader can greatly assist nature by bringing different types of organic matter together to enhance the composting process. By controlling the location, the homesteader can preserve the quality of the finished product. There is no best way to compost; the homesteader does the best with what is available.

Homesteaders are fortunate, because most have the three main ingredients for high-quality compost near at hand: organic refuse

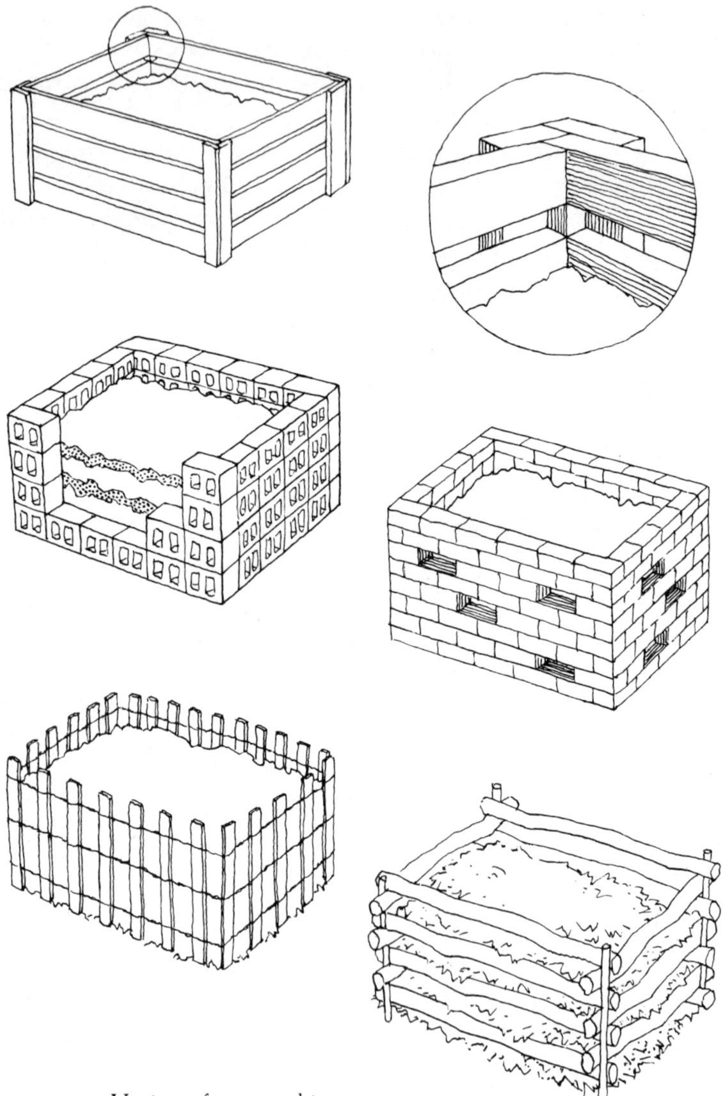

Variety of compost bins

(leaves, straw, garden vines), kitchen refuse (except for fats and meats, which produce a very disagreeable odor while decaying), and manure. In addition, many homesteaders purchase or grow red earthworms, which increase the speed of composting as well as the nutrient value of the finished product. The ingredients are mixed together (do not use wood shavings or sawdust with worms, for it is hazardous to their health).

The highest-quality compost is achieved by providing the optimum amount of mixing, moisture, and nutrients; the higher the quality of compost, the greater the outlay of time, effort, and money. The ideal setup is a number of covered compost bins on cement slabs. Each bin is a four-foot cube, and the material is mixed periodically by moving it from bin to bin in an orderly progression. A more economical system has livestock or people spread the organic material directly onto the land; the material composts directly into the soil.

If you use compost bins, locate them close to the greatest source of raw material, which will probably be manure. They should also be as far away from the house and as far out of sight as possible. The sides should permit plenty of ventilation; they can be constructed out of staggered cement blocks, boards, or saplings arranged like a rail fence. Cover the bins with plastic panels, wood, tar paper, or polyethylene. Even a straw or hay mound heaped high in the middle can be used. This can become the first layer of the next compost bin. If the bins are not constructed on cement they should be moved and the ground underneath planted. There is no sense losing this valuable composted ground.

Whatever composting system you use, the product is like money in the bank. The more you invest, the greater the dividends.

FAT

After butchering most animals, there will be a surplus of fat that can be put to good use. For example, chicken fat is used as a base for making chicken broth; freeze it in small bags and use it as needed.

Goose fat may be melted and used to treat your work boots—it is one of the best waterproofers available—and is also a tasty substitute for butter on popcorn. Once the fat has been rendered, it may be stored in the refrigerator for up to a year.

Salt pork is made from either the choice pork fat cuts from the back or other fat scraps. Cut the fat into 1-pound or $\frac{1}{2}$-pound chunks, making sure you discard any meat. Fill a crock with enough coarse salt to fully cover the fat pieces without crowding. Pack your fat chunks in the salt with none of the pieces touching the sides of the crock or each other. Store the crock in a cool, dry place and in about six weeks the fat will be ready for use as salt pork.

Lard for cooking and baking is rendered pork fat. To make lard, remove any chunks of meat on the fat and cut or grind it into small pieces. Place the fat in a large pan in the oven and set the temperature at 225°F (116°C); fat renders more quickly at a higher temperature, but it will turn a golden brown color instead of the snowy white you want. When well rendered, strain the liquid lard through several layers of cheesecloth to remove the remaining pieces of fat, which are called cracklings. Pour the lard into coffee cans or small crocks, let it cool and then cover. The lard may be kept in the refrigerator for a year. Place the cracklings on a screen over a pan in a 225°F (116°C) oven for a couple of hours. When the cracklings are dry you will have a very tasty snack that will last up to a week in the refrigerator.

Soap is a very worthwhile use of fats. It takes a little time to make, but you can easily produce nine pounds of phosphate- and detergent-free soap for less than seventy-five cents. The type of animal fat affects the hardness of the finished product. Poultry fat produces an objectionably soft soap; sheep tallow produces an excellent soft saddlesoap; pork fat, the most common soap base, is relatively hard; beef tallow is harder than pork fat, and goat tallow makes a very hard soap. You may combine different fats.

First render the fat, either in the oven or in a kettle on top of the stove, keeping the temperature at 225°F (116°C) to maintain a white color. When all the fat is melted, strain it through cheesecloth. Next, clean the fat by bringing it to a boil with an equal amount of water.

Remove it from the heat and chill. The clean fat rises to the top as it cools; when it hardens enough to be removed from the water, turn it upside down and scrape off any impurities.

To make 9 pounds of soap, you need 13 cups of clean fat and a 13-ounce can of lye. Slowly add lye to $2\frac{1}{2}$ pints of cold water in an iron kettle (never use glass or aluminum, because lye generates heat). Always do this outside to disperse caustic fumes. Stir constantly with a wooden spoon to dissolve the lye. The fat and the lye solution should cool to the following temperatures:

Pure lard, 85°F (30°C); lye solution, 75°F (24°C)
$\frac{1}{2}$ lard–$\frac{1}{2}$ tallow, 110°F (42°C); lye solution, 85°F (30°C)
Pure tallow, 130°F (55°C); lye solution, 95°F (35°C)

Slowly add the cooled fat to the lye solution and stir constantly to keep the solution from separating; you may use an egg beater. Stir until it becomes thick as honey, which takes about twenty minutes. Pour the soap into a wood or cardboard box lined with a damp cotton cloth or brown paper. Cover with an old clean blanket to retain the heat, and let it stand for twenty-four hours. Remove the soap and cut into bars with either a knife or a string. Allow the lard to dry in an open area with relatively even temperature for two weeks.

You might vary the soap by adding $\frac{1}{4}$ to $\frac{1}{2}$ cup borax or sugar dissolved in 1 cup of water to either the lye solution or the cooling fat to improve the soap's sudsing ability. Laundry soap can be made by letting the soap age for three or four days and then grating it. Dry your flakes slowly in the oven at the lowest temperature, stirring occasionally. Dishwashing soap can be made by shaving one pound of hard soap and boiling it slowly with 1 gallon of water until the soap is well dissolved. Store in a covered container.

FEATHERS, HIDES, AND WOOL

Feathers, hides, and wool are additional sources of income for the homesteader. Selling them does not generally provide an income sufficient to justify the time involved for a single homestead, but it may be a worthwhile cooperative venture with other homesteaders.

Down from both geese and ducks must be dried before it is stored (see Chapter 8). If you are going to use the feathers yourself, put them in the freezer overnight to keep them from flying all over the house. Homemade down jackets, pillows, and quilts are highly prized items.

If you want to sell the feathers, locate a local buyer or contact a national company, some of which are listed below:

NORTH AMERICAN FEATHER COMPANY
Box 1732
Grand Rapids, MI 49501

PILLOWTEX CORPORATION
1815 North Market Street
Dallas, TX 75202

W.W. SWALEF AND SONS COMPANY
P.O. Box 55743
Fresno, CA 93755

YORK FEATHER COMPANY
10 Evergreen Avenue
Brooklyn, NY 11205

Rabbit pelts are the most valuable to commercial concerns when they are taken from mature rabbits; immature pelts may be marketed locally or put to home use.

To ready hides without tanning, rinse them in cold water and tack them to a board or a pelt stretcher. Scrape any fat or flesh off of the hide and dry the pelt thoroughly. If you are using the pelt yourself, mark off your pattern and cut the pelt with a razor. A furrier's needle is the easiest to use for sewing pelts.

If you want to tan the hides you may have it done commercially or do it yourself. Home tanning uses the following recipe:

2 oz. sulfuric acid	5-gallon nonmetallic container
2 lb. salt	2 gallons cool water

Stir the salt into the water and then slowly add the acid by dribbling it into the water along the side of the container. Never add the water to the acid or let the acid splash; it is caustic before diluted. Stir the solution with a wooden spoon. The solution should be about 70°F (20°C), as higher temperatures can damage the pelts and lower temperatures slow up the tanning process.

Wash the pelt in warm water and detergent and gently squeeze out the water; do not wring. Place the washed hide in the tanning solution and stir. Weight pelts down in the solution with a clean weight (a stone will do) for three to five days, depending upon the size and numbers of pelts. Do not worry about leaving them in too long; pelts can remain in the solution for months without damage, as long as they are stirred occasionally.

When the pelts have finished soaking, take them out and wash them again in warm soapy water, rinse in cold water, and squeeze out the excess water. Remove any flesh and fat remaining on the hide with careful scraping motions, wash the pelt again and rinse. Then put the pelts back into the tanning solution for at least another week.

After removing the pelts from the tanning solution for the final time, again wash, rinse and squeeze the pelts. While the pelt is still damp, gently pull and stretch small areas of the hide in different directions. The hide will turn from brown and stiff to white and soft with this step. When the entire piece is softened, let the pelt finish drying and you have tanned the hide.

Wool may be sold to a shearer or to handspinners, shipped to a mill, or used at home. Handspinner clubs can be contacted through the county extension service or sheep organizations in your area. If you want to have wool converted to yarn for your own use, one place you may send it to is *Bartlettyarns, Inc., Harmony, ME 04942*. It takes about two pounds of wool in the grease (as it comes from the sheep) to produce one pound of finished yarn. It takes about fourteen pounds of wool in the greese to produce a twin-size blanket.

To prepare the wool for storing, selling, or shipping, spread the fleece out on a clean surface with the skin side down. Fold the side edges in toward the middle and fold in the neck edge. Roll from the tail end and make a compact roll. Tie the roll tightly at both ends and store in a burlap bag or feed sack that allows the fleece to breathe.

If you plan to spin or make batts (a mass of cotton fibers) from your own wool, you will have to wash and card it. Before washing the wool, remove as much of the chaff and tags from it as possible. Wash the wool in warm water and detergent and rinse in cold water twice (sudden temperature changes can damage wool). We found that young children are excellent wool washers, as they seem to resist Mom's tendency to overwash, thus removing the oil and making the wool unspinnable. Once the wool is washed, dry it in the sun, if possible. The dry wool is then carded; cards may be purchased at farm-supply stores.

FENCING

All homesteads with animals require fencing. What type of fencing you use depends on size, time, availability of materials, and budget.

Wood fences and posts should be made from seasoned wood and treated with a preservative to increase longevity. The best preservative is creosote, but it is also the most expensive. Motor oil is a good alternative, by itself or in combination with creosote; you may use used motor oil. Soak the wood in the preservative rather than paint it on.

Set the posts deep; line posts about $2\frac{1}{2}$ feet down, and corner posts about $3\frac{1}{2}$ feet deep. A post-hole digger is a helpful tool to have on hand.

Woven wire fencing comes in different heights, widths (distance between vertical stays), and weight. The style of woven wire is indicated by a number; the first digit or two indicates the number of horizontal lines, the last two digits indicate the height. Thus, an 1155 fence is fifty-five inches high with eleven horizontal wires, with five inches between horizontal lines. Your choice of stay distance will be either six or twelve inches.

The weight of the woven wire depends on the gauge; the lower the number, the heavier and more expensive the wire. Medium weight is best for livestock. Woven wire is strung with a rod ($16\frac{1}{2}$ feet) between line posts. Fasten the wire to the posts on the inside of the fence.

Barbed-wire fencing can be used by itself or in conjunction with other types of fencing. You have a choice of gauge and number of barbs. Choose a heavier gauge and two-point barbs for all stock except cattle, which require four-point barbs. Line posts are set at a distance of a rod apart. An all-barbed-wire fence should not be used if you are enclosing a small area, nor for horses.

Electric fencing can be used alone or in conjunction with other fencing; our experience has been that the most effective fencing system is a combination of woven wire and electric fencing, a strand of electric wire strung above and inside the woven wire.

The charge for electric wire is transmitted through the wire from a control box and is strong enough to give a good jolt without causing harm to stock or humans. Conventional electric wire requires that insulators be placed at each corner and on line posts.

An exceptionally fine electric fence system is charged by the New Zealand fence controller. This unit is more expensive to buy than the standard one, but the added benefits justify the difference in cost. The New Zealand controller charges longer fence lines, is less susceptible to shorting out on grass or weeds, and allows for greater distance between fence posts.

Grounding wire fences is an important preventative measure; current from lightning can travel as far as two miles on an ungrounded wire. Wire fencing can be grounded by setting galvanized steel posts into the ground at intervals of 150 feet. You may successfully ground a wire fence by setting a 5-foot-long $\frac{1}{2}$- or $\frac{3}{4}$-inch steel pipe into the ground along wooden posts at 150-feet intervals. The pipe may be fastened to the posts with pipe straps so it touches all the fence wires.

BACK AT THE FARM

GARDEN CROPS FOR ANIMAL FEED

Overproductive gardens can be the norm for homesteaders, but garden surplus and thinnings can successfully be used to stretch your feed dollar as long as you use them as dietary supplements. Remember to add your succulent castoffs gradually to your animal's diet to avoid digestive problems. Tubers and stalks fed to animals should be chopped to avoid choking. The following chart indicates the type of vegetable preferred by various animals.

Vegetable	Edible Part	Rabbits	Goats and Sheep	Pigs	Fowl	Cows	Horses
Beans (wax, green)	Pods, vines		x	x			
Cabbage		in small amounts	x	x	x		
Carrots	Root	x	x	x	x	x	x
Comfrey	Leaves	x	x	x	x	x	x
Corn (sweet, Indian, field)	Husks, stalks, cob	x	x	x	x	x	x
Beets	Roots, greens		tops	x	tops	x	x
Greens (chard, kale, spinach)	Leaves		x	x	x	x	x
Kohlrabi	Fruit, top		x	x			
Parsnips	Root, top		x	x		tops	
Peas	Pods, vines		x	x			
Potatoes	Fruit, split open		x	x	x	x	
Squash, Pumpkin	Fruit, split open		x	x	x	x	
Rutabagas	Fruit			x			
Tomatoes	Fruit			x	x		

GESTATION AND INCUBATION PERIODS

The following table provides the average gestation or incubation periods for animals commonly found on a homestead:

Animals	Number of Days
Cat	63–65
Chicken	21
Cow	279–290
Dog	58–63
Duck	28
Goat	145–155
Goose	28–34
Guinea Fowl	28
Horse	330–345
Pig	112–115
Rabbit	30–34
Sheep	145–155
Turkey	28

MEAT VALUES

The U.S. Department of Agriculture provides the following values of various meat:

Meat	Fat(%)	Moisture(%)	Protein(%)	Calories per Pound
Beef	28	55	16.3	1,440
Chicken	11	67.6	20	810
Duck	28.6	54.3	16	1,470
Goose	31.5	51.6	16.4	1,580
Lamb (*medium fat*)	27.7	55.8	15.7	1,420
Pork (*medium fat*)	45	42	11.9	2,050
Rabbit	10.2	27.9	20.8	795
Turkey (*medium fat*)	22.2	58.3	20.1	1,190
Veal (*medium fat*)	14	66	18.8	910

MEDICATION

Occasionally you will have to administer worming or other medications to your animals. Trying to stuff a large pill down the throat of a horse or a 250-pound ram is a bit different than slipping Rover a vitamin, but fortunately there are some inexpensive instruments to assist you.

A bolus is a large pill for farm animals, and it can most easily be administered with the aid of a bolus or balling gun. If the animal's size allows, back it into a corner and staddle it, facing its head. Put the bolus in the gun and insert it into the mouth behind the tongue, using your free hand to wedge the mouth open. When in position, pop the bolus down the animal's throat. Do not let the animal go until you are sure the medication has been swallowed. The bolus will go down more easily if it has been coated with mineral or vegetable oil.

Liquid medications can be administered with a dose syringe; the procedure is called drenching. Straddle the animal to be drenched, facing forward. Insert the nozzle of the syringe into the animal's mouth, tilting the head upward slightly. Do not tip the head too high or you may force the liquid into the animal's lungs. Administer the liquid slowly, allowing the animal to swallow frequently.

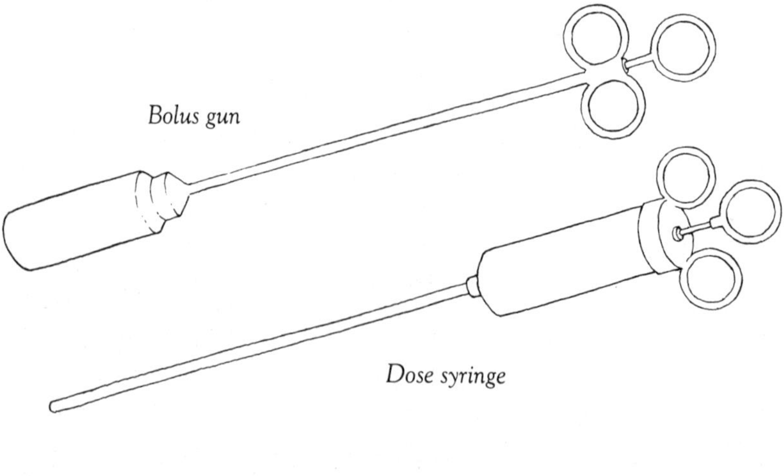

Bolus gun

Dose syringe

Injections may be required, and different medications require different types of injections. Learn this skill from a veterinarian or other knowledgeable persons.

NEIGHBORING

The term *neighbor* not only refers to people who live close to each other, but also to those who have something in common. In homesteading, the verb *neighboring* has come to mean a relationship of mutual assistance. Self-sufficiency does not mean isolation. Homesteaders understand well the importance of give-and-take with others. The relationship is beneficial materially and psychologically. We have found that exchanging recipes, tools, and work, successes, hardships, and encouragement has been one of the greatest benefits of homesteading. There is no place for vanity or competition in homestead neighboring.

Perhaps the most important neighbor and certainly one of the first to contact in a new area is your county extension agent, a professional neighbor. Every county has a cooperative extension service staffed by a group of highly trained people supported by a cadre of scientists and other specialists.

County agents are the gatekeepers of a storehouse of knowledge: from how to locate, purchase, and finance your homestead to how to process your produce. The knowledge comes in three forms: pamphlets, the county agent and his or her support services, and referrals (both professionals and other homesteaders). County agents are part scholar, part scientist, part teacher, and three parts neighbor. They are found in county extension offices along with home economists and 4-H Club directors. Behind this group of direct consultants are the educators and scientists in the state land-grant universities and U.S. Department of Agricultural experiment stations.

This army of neighbors has been producing and sharing information through federally established land-grant schools since 1862. By 1914 it was found that agricultural education solely on college campuses was inadequate, and Congress created the Cooperative Extension Service, a partnership between the state land-grant universities and the U.S. Department of Agriculture.

County agents are unique neighbors. They are professional consultants. They are paid by you and other taxpayers to help you solve problems; they will not solve your problems for you. They will help you identify your problems, determine your objectives, locate information to establish alternatives, and be your mentor as you choose the most feasible alternative.

Also, your county agent is a conduit with other homesteading neighbors. A formal way is through 4-H Clubs, where you can affiliate with parents of other young homesteaders. You also may wish to join a homesteaders' organization. If you are not a joiner, you may wish to meet others who raise a particular kind of animal; your county agent can put you in touch with these neighbors.

Homesteading neighbors are vitally important. We all need other homesteaders to test reality: "Are we crazy living this way?" "Oh, you feel that way too! Good, I'm not crazy, we're just different."

PEST CONTROL

Pest control is a continuous battle on the homestead. Lice and mites can be controlled by frequently checking your animals and applying an appropriate compound when the problem is detected. Frequent grooming also controls lice and mites.

Mice and rats can be a particularly offensive problem. These repulsive rodents are dangerous because they can contaminate feed and transmit disease. Young poultry and waterfowl are easy prey for rats. Unfortunately, most rodenticides are also poisonous to farm animals and must be used with extreme caution, if at all. Traps can control rodents. The best preventative measure is to use metal barrels with tight-fitting lids for grain storage. Barn cats and owls are also a big help.

Poultry and waterfowl can attract skunks, opossums, raccoons, and foxes. A good fencing system, closing up your birds in a shelter at night, and a good farm dog should just about eliminate this problem.

The fly season provides the conscientious homesteader with a real challenge. Some control measures that can be regularly executed include the following:

- Clean manure frequently.
- Spread manure thinly outdoors, so that the fly eggs and larvae can be killed by drying.
- Do not allow wet litter, old wet hay or straw, or the compost heap to be in or near the barn.
- Eliminate improper drainage areas near the barn or in the pasture area.

If all else fails, you may want to use an insecticide. **Use insecticides cautiously and only as the directions on the label indicate.** Do not use Diazinon or any of the chlorinated hydrocarbon compounds on your homestead if you have dairy animals. Some chemicals, such as malathion, can be applied to the walls and ceiling of your barn, either directly or added to whitewash. Use these chemicals only as indicated. Fly strips and baits may be used, but must be kept out of reach of your animals.

POISONOUS PLANTS

There are many plants that are poisonous to animals. Check all grazing areas to make sure that none is accessible to your stock. The following list is by no means complete, but indicates the range of plants involved; contact your local agricultural extension agent for a complete listing of poisonous plants indigenous to your area for your type of stock.

Arrowgrass
Bracken fern
Broomcorn
Buckeyes
Buttercups
Castor bean
Chokecherries
(*wild cherry*)
Cocklebur
Corn cockle
Dogbane
Fly-poison
Hemlock
Jessamines (yellow)
Jimson weed
Johnson grass
Laurels
Laurel cherry
Larkspurs
Locoweed
Nightshades
Oaks
Rattlebushes

PRODUCT: FEED CONVERSION RATIO

The following table provides the amount of feed required to produce one unit of produce:

Produce	Amount of Feed	Amount of Product
Beef	20 lb.	1 lb.
Chicken	3 lb.	1 lb.
Eggs	4.5 lb.	1 doz.
Lamb	4 lb.	1 lb.
Milk	2.6 lb.	1 gal.
Pork	6 lb.	1 lb.
Rabbit	4 lb.	1 lb.

SLAUGHTER AGE AND WEIGHT

The following table provides the average age and weight considered desirable for slaughtering various animals:

Breed	Age	Live Weight
Beef	16–18 months	1,000–1,500 lb.
Chicken (broiler)	8–10 weeks	4–6 lb.
Duck	10–16 weeks	6–8 lb.
Goose	16–20 weeks	10–15 lb.
Goat	5–6 months	50–60 lb.
Rabbit	8–10 weeks	4–5 lb.
Sheep	5–6 months	80–100 lb.
Turkey	4+ months	10+ lb.
Veal	8–10 weeks	175–250 lb.

TATTOOING, TAGGING, AND BANDING

Registered and valuable animals should be tattooed, tagged, or banded for identification. Tattooing is done with a tattoo needle,

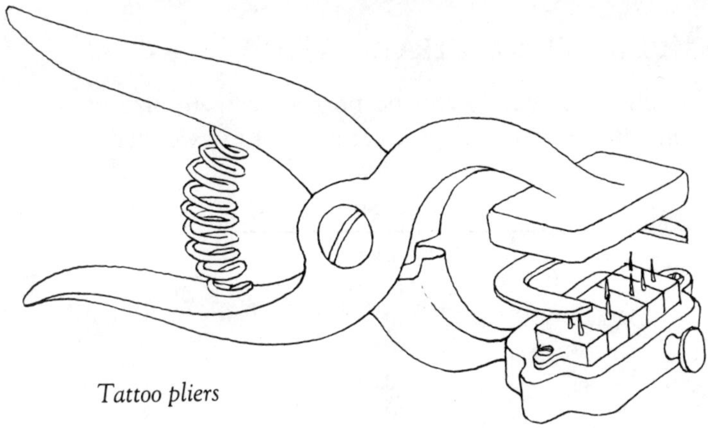

Tattoo pliers

tagging with an ear punch or a self-piercing tag, and banding is placing a band on the legs or wings of birds. All three systems are relatively inexpensive.

To tattoo, place the number or letter die in a tattoo holder and do a test punch on a piece of paper to make sure you have the proper sequence of numbers or letters. Clean the area on the animal to be tattooed with a piece of cotton dipped in alcohol. Apply tattoo ink over the area, then place your tattoo holder in position and puncture the skin with a firm squeeze. Do it quickly to avoid much pain. Remove the tattoo instrument and rub the ink into the holes with your finger. Do not try to wash the excess ink off until the skin has healed over the puncture wounds, which takes about a month.

Tattooing is generally done in the left ear, taking care to avoid ear ridges. Some animals, such as LaManchas and Nubians, are tattooed in the tail webbing. Use $\frac{1}{4}$-inch die for goats and sheep and $\frac{3}{8}$-inch die for rabbits.

Ear tagging is recommended only for cows. Goats, sheep and rabbits are not ear tagged because of the chance of catching the tag on a fence, wire, or brush and ripping the ear.

Banding birds may be done to provide a means of quick identification of gender and age. The bands are quite inexpensive and easily fitted on the bird's leg or wing.

TEMPERATURE CHART

The following table gives the normal temperatures for various animals; allow for a one-degree variance for individual difference:

Animal	Fahrenheit	Centigrade
Cat	101.5	38.5
Cow	101.5	38.5
Dog	102	39
Goat	102	39
Horse	100.5	38
Pig	102	39
Rabbit	102.5	39.3
Sheep	102.2	39

WHITEWASH

A barn painted on the inside not only looks nicer, but also goes a long way toward maintaining higher sanitary standards. Do not use lead-based paints. The cheapest method is to use good old-fashioned homemade whitewash.

Surface preparation means scraping the walls and/or brushing with a stiff brush. Before you apply the whitewash, the surface should be dampened so that the whitewash will dry slowly and not chalk up.

Refined hydrated lime is the best lime to use to mix up the paste for whitewash. It takes approximately fifty pounds of hydrated lime slaked with six gallons of water to make eight gallons of lime paste. Slaked lime alone can be used for whitewashing, but your efforts will last longer if you add other materials to the mix. The simplest formula is to dissolve fifteen pounds of salt in about five gallons of water and add this solution to the lime paste. Mix thoroughly and add water to achieve desired consistency.

CONCLUSION

THIS book makes it clear that homesteading is a way of life, not a hobby. It requires knowledge, time, effort, and above all a commitment to the goal of self-sufficiency. Farming is generally accepted and societal demands are adjusted accordingly, but the homesteader must straddle a narrow line: a farmer who misses a meeting because he must get the hay in is excused, but a homesteader who is late for class because his ewe had difficulty lambing is problematic. Your commitment makes the difference.

Homesteading involves a great deal of hard, inflexible work that frequently is filled with frustration and disappointment. Crops fail, animals die, projects get interrupted, animals need to be fed even during blizzards when you all have terrible colds. You will not be free to take off on spur-of-the-moment trips. You will find yourselves processing vegetables, complete with steaming water bath, when it is 100°F (38°C) outside with high humidity and you would like nothing better than to shuck the whole thing and take the kids swimming. Friends may offer you tickets to a ballgame or concert, but you will have to miss it because you have to mend fences or dung out the barn.

CONCLUSION

The rewards are there, too, and we believe they outweigh the negatives. There are the successful crops, the new generation of young animals, the freezer and pantry full of food, and the absence of fear when yet another food additive is found to be carcinogenic. There are those inspirational trips to the grocery store where you view the spiraling food costs with a shudder of relief, realizing that your greatest food problem is whether you can eat enough from the freezer before the next crop needs to go into it. There is the pride your children feel when they announce in school that *they* bring home the bacon.

We can promise that your homesteading chores will not be boring or trite; they will help increase your self-discipline, and you probably will not need to jog for your physical health. You will feel very much alive! Homesteading chores put you in touch with the soil, the weather, the seasons, and all living things. The work spans centuries, cultures, and continents, and your goals are and have been universally shared. You are in touch with your fellow beings. The rewards are great, and as for the hard work, perhaps Sir John Lubbock expressed it best: "The idle man does not know what it is to rest. Hard work, moreover, not only tends to give us rest for the body, but, what is even more important, peace to the mind."

BIBLIOGRAPHY

THERE are many additional sources of general and specific information available to you. Your local agricultural extension office can provide you with a list of pertinent government bulletins, many of which are available free. We encourage you to make use of these most helpful publications.

In addition, there are many books and magazines designed to assist homesteaders in various undertakings. Below we provide a listing of magazines, books and supply houses that we have found helpful to us. It is by no means complete, but we hope it is sufficient to help you get started in a number of specialized directions.

MAGAZINES

BLAIR & KETCHUM'S COUNTRY JOURNAL/139 Main Street, Brattleboro, VT 05301

COUNTRYSIDE/312 Portland Road, Waterloo, WI 53594

DAIRY GOAT JOURNAL/P.O. Box 1908 c-4, Scottsdale, AZ 85252

DRAFT HORSE JOURNAL/Route 3, Waverly, IA 50677

THE EVENER/Putney, VT 05346

BIBLIOGRAPHY

MOTHER EARTH NEWS/P.O. Box 70, Henderson, NC 28739

ORGANIC GARDENING/33 East Minor Street, Emmaus, PA 18049

THE SHEPARD/R.D. 1 Box 67, Sheffield, MA 01257

BOOKS

Belanger, Jerome. *The Homesteader's Handbook to Raising Small Livestock.* Rodale Press, 1976.

Bennett, Robert. *Raising Rabbits the Modern Way.* Garden Way Publishing Co., 1975.

Hertzberg, Ruth, Beatrice Vaughan, and Janet Greene. *Putting Food By.* Stephen Greene Press, 1975.

Hobson, Phyllis. *Making Homemade Cheeses and Butter.* Garden Way Publishing Co., 1975.

Klein, G. T. *Starting Right with Poultry.* Garden Way Publishing Co., 1975.

Langer, Richard. *Grow It!* Avon Books, 1972.

Lockwood, Guy. *Animal Husbandry and Veterinary Care.* White Mountain Publishing Co., 1977.

Luttman, Rick and Gail. *Chickens in Your Backyard.* Rodale Press, 1975.

Merck & Co. ed. *The Merck Veterinary Manual.* Merck & Co., 1961.

Morrison, Frank. *Feeds and Feeding.* Morrison Publishing Co., 1961.

Sheraw, Darrel. *Successful Duck and Goose Raising.* Stromberg Publishing Co., 1975.

Simmons, Paula. *Raising Sheep the Modern Way.* Garden Way Publishing Co., 1976.

Spaulding, C. E. *A Veterinary Guide for Animal Owners.* Rodale Press, 1976.

Szyketka, Walter, ed. *Public Works: A Handbook for Self-reliant Living.* Links Books, 1974.

BIBLIOGRAPHY

SUPPLY HOUSES

Horses

Phil Stanton/Wild as the Wind, Upton, MA 01568

Smucker Harness Shop/Box 48, Churchtown, PA 17510

Village Harness Shop/R.F.D. 1, Ronks, PA 17572

David Fisher/Fisher's Carriage Shop, R.F.D. 3, New Holland, PA 17557

Rabbits

Bass Equipment Company/Box 352, Monett, MO 65708

Valentine Equipment Company/9706 S. Industrial Drive, Bridgeview, IL 60455

Goats

American Supply House/Box 1114, Columbia, MO 65205

Hoegger's/Box 490099, College Park, GA 30349

Poultry and Fowl

Mt. Healthy Hatcheries, Inc./Mt. Healthy, OH 45231

Stromberg's/Pine River, MN 56474

Sheep

Sheepman Supply CO./P.O. Box 100, Barboursville, VA 22923

General Supplies and Books

Countryside Catalogue/312 Portland Road, Waterloo, WI 53594

Cumberland General Store/Route 3, Box 479, Crossville, TN 38555

Garden Way/Charlotte, VT 05445

Glen-Bels Country Store/Route 5, Crossville, TN, 38555

INDEX

INDEX

INDEX

INDEX

INDEX